TECHNOLOGY IN A GLOBALIZING WORLD

NONLINEAR SYSTEMS

CHAOS, ADVANCED CONTROL AND APPLICATION PERSPECTIVES

TECHNOLOGY IN A GLOBALIZING WORLD

Multidisciplinary Science and Advanced Technologies
Kaushik Pal, PhD, Fernando Gomes, PhD and Thinakaran Narayanan (Editors)
2021. ISBN: 978-1-53618-959-9 (Hardcover)
2021. ISBN: 978-1-53619-198-1 (eBook)

Issues with Facial Recognition Technology
Warren Lambert (Editor)
2020. ISBN: 978-1-53618-973-5 (Hardcover)
2020. ISBN: 978-1-53619-000-7 (eBook)

The Pulp and Paper Industry: Production, Management and Technology
Mette P. Kristensen (Editor)
2020. ISBN: 978-1-53617-919-4 (eBook)

A Closer Look at Educational Technology
Maria A. Clausen (Editor)
2019. ISBN: 978-1-53616-215-8 (Softcover)
2019. ISBN: 978-1-53616-216-5 (eBook)

More information about this series can be found at
https://novapublishers.com/shop/a-closer-look-at-educational-technology/

TECHNOLOGY IN A GLOBALIZING WORLD

NONLINEAR SYSTEMS

CHAOS, ADVANCED CONTROL AND APPLICATION PERSPECTIVES

PIYUSH PRATAP SINGH

Copyright © 2022 by Nova Science Publishers, Inc.
https://doi.org/10.52305/GEAW7683

All rights reserved. No part of this book may be reproduced, stored in a retrieval system or transmitted in any form or by any means: electronic, electrostatic, magnetic, tape, mechanical photocopying, recording or otherwise without the written permission of the Publisher.

We have partnered with Copyright Clearance Center to make it easy for you to obtain permissions to reuse content from this publication. Simply navigate to this publication's page on Nova's website and locate the "Get Permission" button below the title description. This button is linked directly to the title's permission page on copyright.com. Alternatively, you can visit copyright.com and search by title, ISBN, or ISSN.

For further questions about using the service on copyright.com, please contact:
Copyright Clearance Center
Phone: +1-(978) 750-8400   Fax: +1-(978) 750-4470   E-mail: info@copyright.com.

NOTICE TO THE READER

The Publisher has taken reasonable care in the preparation of this book, but makes no expressed or implied warranty of any kind and assumes no responsibility for any errors or omissions. No liability is assumed for incidental or consequential damages in connection with or arising out of information contained in this book. The Publisher shall not be liable for any special, consequential, or exemplary damages resulting, in whole or in part, from the readers' use of, or reliance upon, this material. Any parts of this book based on government reports are so indicated and copyright is claimed for those parts to the extent applicable to compilations of such works.

Independent verification should be sought for any data, advice or recommendations contained in this book. In addition, no responsibility is assumed by the Publisher for any injury and/or damage to persons or property arising from any methods, products, instructions, ideas or otherwise contained in this publication.

This publication is designed to provide accurate and authoritative information with regard to the subject matter covered herein. It is sold with the clear understanding that the Publisher is not engaged in rendering legal or any other professional services. If legal or any other expert assistance is required, the services of a competent person should be sought. FROM A DECLARATION OF PARTICIPANTS JOINTLY ADOPTED BY A COMMITTEE OF THE AMERICAN BAR ASSOCIATION AND A COMMITTEE OF PUBLISHERS.

Additional color graphics may be available in the e-book version of this book.

Library of Congress Cataloging-in-Publication Data

ISBN: 978-1-68507-660-3

Published by Nova Science Publishers, Inc. † New York

*Dedicated to the
lovely sweetheart son
Prakrit
and
to that
chaotic
never to be understood
chameleon
provoking diligent
and
absolutely necessary
person
the gentle reader*

CONTENTS

Preface ix

Chapter 1   **Real and Complex State Variables Based Autonomous Chaotic Systems** 1
Sugandha and Piyush Pratap Singh

Chapter 2   **Projective Synchronisation of Complex Valued Hyperchaotic Systems and Encryption Techniques in Secure Communications** 23
Sugandha and Piyush Pratap Singh

Chapter 3   **Chaos Control in Electric Vehicle Using Cascaded PI and PID Sliding Mode Control Techniques** 49
Ranjan Kumar, Nathaneal Kharlukhi, Md. Tanweer Alam and Piyush Pratap Singh

Chapter 4   **Chaos Control in Power System Using Higher Order Sliding Mode Control** 83
Manish Kumar and Piyush Pratap Singh

Chapter 5   **Chaos Control in the Rotational Machine with Centrifugal Flyball Governor System Using Backstepping Sliding Mode Control** 113
Piyush Pratap Singh and Binoy Krishna Roy

About the Editor 133

Index 135

PREFACE

The book covers different subjects in the field of nonlinear dynamics, especially applications and investigation of chaos and chaotic systems in electrical engineering, information technology, communication engineering and mechanical engineering. This book is suitable as a textbook at the graduate or advanced undergraduate level and will appeal to postgraduate-level students and young researchers in different fields.

This book provides technological advancement in nonlinear dynamics and chaos and explores the fields of communication, electric vehicles, power systems and rotational machines with centrifugal flyball governor system. An autonomous chaotic system is explored with real and complex state variables; their projective synchronisation is reported with application to secure communication. Secure communication is achieved using Masking-Modulation and Diffie-Hellman Key Exchange encryption techniques. Further, electric vehicles are the necessity of upcoming trends. To optimize the control performance of the permanent-magnet synchronous motor with different disturbances and uncertainties, a nonlinear control for the permanent-magnet synchronous motor using sliding-mode control is reported and Cascaded PI sliding mode control technique is explored to control the chaotic behaviour in electric vehicles. Chaos behaviour is explored in power systems and its control is presented using higher order sliding mode control. Comparative performances are analysed followed by control of chaos in the Rotational Machine with Centrifugal Flyball Governor system where chaos is controlled using recursive backstepping sliding mode control. All the simulations are carried out in the MATLAB environment and reveal successful achievement of the objectives.

Researchers from academia and industry, who are working in the research areas Nonlinear Dynamical Systems & Chaos, Electrical Engineering, Computer Science Engineering, Information Technology, Communication Engineering and Mechanical Engineering may be principal audiences. Also, the book will be helpful for (i) graduate or advanced undergraduate level students as a textbook or major reference book for courses such as electrical circuits, nonlinear dynamical systems, mathematical modelling, computational science, numerical simulation, and many others and (ii) postgraduate level students and young researchers in the following fields: Communication Engineering; Computer Science; Electrical and Electronic Engineering; Mechanical Engineering; Engineering Mathematics; Computational Physics.

In: Nonlinear Systems
Editor: P. P. Singh

ISBN: 978-1-68507-660-3
© 2022 Nova Science Publishers, Inc.

Chapter 1

REAL AND COMPLEX STATE VARIABLES BASED AUTONOMOUS CHAOTIC SYSTEMS

Sugandha[*] and Piyush Pratap Singh[†]
Department of Electrical Engineering,
National Institute of Technology Meghalaya,
Shillong, Meghalaya, India

Abstract

In this chapter, a novel three dimensional (3D) chaotic system based on the real state variables is proposed and analysed. Further, a complex state variables based dynamics is derived from the proposed 3D chaotic system and analysed. The proposed and derived systems are then examined for the various dynamic properties such as, its equilibrium points, symmetry and invariance, dissipativity and Lyapunov exponent etc. Different qualitative and quantitative tools such as phase plane, time series, Lyapunov exponents, Lyapunov spectrum, Lyapunov dimension are used to evidence the different dynamic behaviours of the proposed 6D hyperchaotic system. Simulation is presented in MATLAB environment to demonstrate the effective verification of the theoretical approach. Simulation results oblige that the objectives are achieved successfully.

---

[*]Corresponding Author's E-mail: itz.sugandha11@gmail.com.
[†]Corresponding Author's E-mail: piyushpratapsingh@gmail.com.

1. INTRODUCTION

Chaos theory, as a definition, is the study of complex non-linear dynamic system. It's a branch of mathematics that deals with deterministic processes. In the normal language, the term *chaos* is considered as a place of total disorder and confusion. There has been a literal explosion of scientific interest about the term *chaos* and its control. Chaos behaviour is different from the random or noise behaviour. In scientific literature, the word *chaos* means anything which appears to be disordered, aperiodic long-time behaviour and deterministic in nature. The system which follows these properties is known as *chaotic system*. As an example, the famous pinball machine is a classic example of chaotic system and termed as Lorenz's chaotic system [1]. Anyone who has ever tried to release a pinball at the right speed to repeat the desired trajectory knows that it can't be done except on seemingly random occasions. It is not that the pinball's course is unpredictable, it's just that the resulting trajectory is highly sensitive to minor changes in the initial conditions.

Recent developments in the field of chaos theory have made its availability as a spreading signal in a variety of disciplines such as communication: cryptography and spread spectrum modulation due to its characteristics [2]. The chaos behaviour is used to provide security when information is being transmitted via Telecommunication technology. Security of communication depends upon the complex behaviour of the transmitted signal. In this process, data encoding is based on the properties of chaotic dynamics such as deterministic in nature, complex behaviour and spread spectrum. As a deterministic nature, the decoding of the data or information signal is possible. In practice, the implementation of chaos-based communication requires chaotic system or oscillator synchronisation at both the transmitting and receiving ends. The synchronisation method is now commonly used to encrypt and decrypt messages [3]. Information security is one of the important issues now-a-days. The encryption and decryption method should be arranged as to prevent interception and becomes the primary motivation to this work.

1.1. Chaotic System

Chaotic system has complex behaviour which shows sensitivity to initial conditions, such as an economy, a stock market, weather etc. in such systems any small uncertainty in the beginning produces rapidly escalating and compounding errors in the prediction of future behaviour. A chaotic system has following properties:

1. Sensitive to initial conditions

2. Topologically mixing

3. Dense aperiodic orbits

4. Deterministic nature

One of the basic examples of a chaotic system is a Rossler's chaotic system [4]. The dynamics of Rossler's chaotic system is given by:

$$\begin{cases} \dot{x} = -y - z \\ \dot{y} = x + ay \\ \dot{z} = b + z(x - c) \end{cases} \quad (1)$$

There are various qualitative and quantitative tools are available to detect the chaos behaviour and discussed in subsection 1.2

1.2. Tools for Detecting Chaos

The graphical representaions such as time series, phase portrait, Lyapunov spectrum, bifurcation diagram and power spectrum are known as qualitative tools. Calculation of Lyapunov exponent and Lyapunov dimension are termed as quantitative tools. Using Rossler's dynamics in (1), the different tools are discussed as follows:

1.2.1. Time Series Plot

The system's state variables are observed in this graphical method and if the behaviour of state variables are irregular or unpredictable, then it is called chaotic behaviour. This is the easiest graphical method. In the observation, one can find other non-chaotic behaviours such as fixed point,

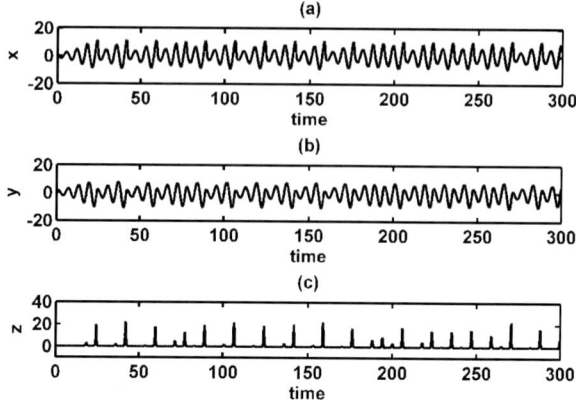

Figure 1. Time series behaviour of Rossler's system (1) shows chaotic behaviour at $a = 0.2$, $b = 0.2$ and $c = 5.7$.

periodic and quasi periodic. It is observed that the system variables of system (1) exhibits irregular or chaotic behaviour as shown in Fig. 1 at the parameter values $a = 0.2$, $b = 0.2$ and $c = 5.7$. Figures 3 & 4 show the periodic and period-2 (quasi-periodic) behaviours at parameter values $c = 2.3$ and 2.9, respectively, and other parameters are fixed.

1.2.2. Phase Portraits

Phase plane is a two-dimensional projection of the phase-space [5]. The phase plane behaviour starting from different initial condition in a single phase plane plot is known as phase portrait. Phase portrait represents each of the state variable's instantaneous state to each other. In a phase portrait, a fixed point solution represents a point, a periodic solution shows a closed curve and chaotic solutions represent distinct curves. For Rossler's system (1), chaotic behaviour is shown in the Fig. 2. The periodic and period-2 (quasi-periodic) behaviours are shown in Figs. 3 and 4, respectively.

1.2.3. Lyapunov Exponents

Lyapunov exponents $(\lambda = 1, 2, ..., n)$ measure the exponential attraction or separation of two neighboring orbits in phase space. An n dimensional system has n Lyapunov exponents. When the largest Lyapunov exponent

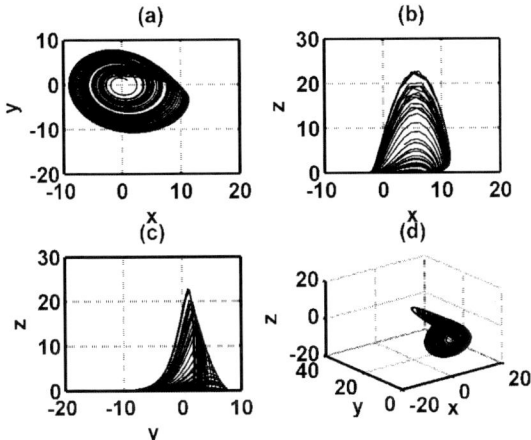

Figure 2. Phase portrait of Rossler's system (1) at $a = 0.2$, $b = 0.2$ and $c = 5.7$.

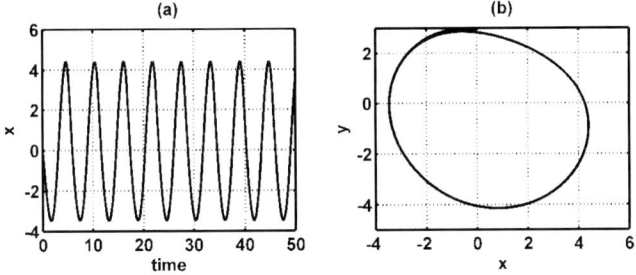

Figure 3. Time series and phase plane plots of Rossler's system (1) shown in (a) and (b), respectively, reflect periodic behaviour at $c = 2.3$.

is negative then the attractors converge in time and system becomes insensitive to initial conditions. When it is positive, then the distance between neighboring orbits increases exponentially and the system shows sensitive dependency on the initial conditions, and it is assumed to be chaotic. For Rossler's system (1), the Lyapunov spectrum is plotted by varying c parameter and keeping the other parameters fixed as shown in Fig. 5. From Fig. 5, it is observed that system (1) exhibits periodic behaviour when $c \in [1, 2.48) \cup (2.92, 3.6] \cup (29.71, 30.58)$, period-2 behaviour

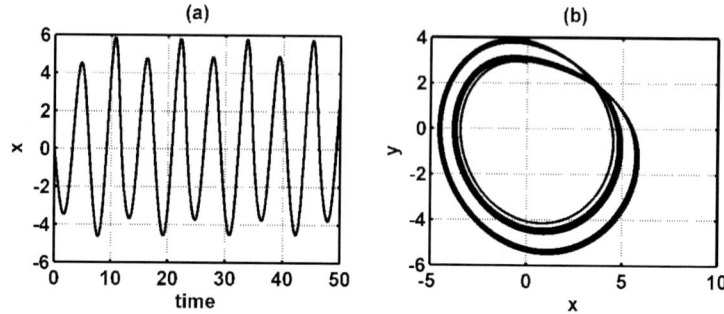

Figure 4. (a) Time series and (b) phase plane plots of Rossler's system (1) shows period-2 behaviour at $c = 2.9$.

when $c \in [2.48, 2.92] \cup (3.6, 4.26) \cup (29.17, 29.71]$ and chaotic behaviour for $c \in [4.26, 29.17] \cup [30.58, 35]$.

1.2.4. Bifurcation Diagram

In differential equation, the change in number of solution depends on the variation in the parameter, is known as bifurcation [5]. It is used to explain specific variation in the behavior of the system under the variation of one or more system parameters, in terms of type and number of solutions. Figure 5 represents the bifurcation diagram of Rossler's system (1). The bifurcation diagram is plotted by varying c parameter and keeping other parameters fixed. Bifurcation diagram shown in Fig. 5 supports the corresponding Lyapunov spectrum behaviours.

1.2.5. Power Spectrum

Chaotic signals are treated as wideband in nature by looking at their frequency spectra. For chaotic behaviour, the power spectra of system is expressed in terms of oscillations with a continuum of frequencies. The single sided power spectrum of $x(t)$ signal of Rossler's system (1) is shown in Fig. 6. Random peaks in the spectrum indicates the chaotic behaviour.

Using the above dynamic analysis tools, the basic dynamical properties of the proposed chaotic system (2) are explored further in the next Section.

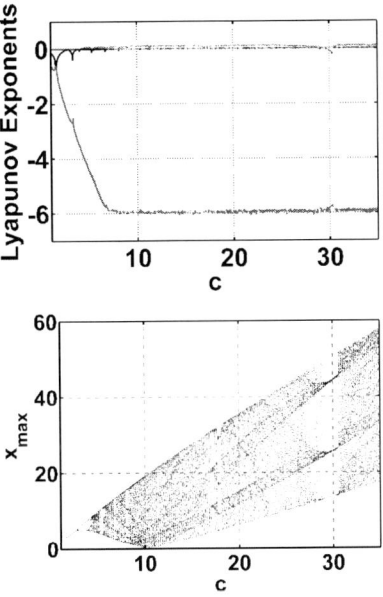

Figure 5. Lyapunov exponent and bifurcation diagram of Rossler's system (1).

2. LITERATURE REVIEW

In 1979, Otto Eberhard Rossler, a German biochemist, had discovered hyperchaos phenomenon with a four dimensional autonomous system for the very first time and known as hyperchaotic Rossler system [4]. Later, many other hyperchaotic systems like Colpitt's, Chen's, Lu's, Lorenz's etc. were also constructed via. anti-control and analysed in [8, 9, 10, 11, 12, 13, 14]. In hyperchaotic system, the exponential divergence between closely adjacent trajectories occurs simultaneously in more than one direction, i.e., technically more than one positive Lyapunov exponents are required or present in dynamical system [15] which makes the behaviour of a hyperchaotic system more complex than the chaotic system. To have more than one positive Lyapunov exponent, the dimension of a chaotic system needs to be increased.

Recently, various research have been explored on the design of hyperchaotic system by using complex state variables in place of real state vari-

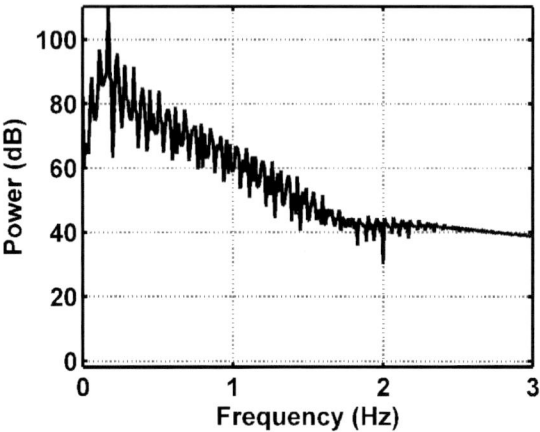

Figure 6. Power spectrum of $x(t)$ state of Rossler's system (1).

able and its synchronisation [16, 17, 18, 19, 20, 21]. Complex state variables based dynamics of the existing chaotic systems such as Lu chaotic system [16, 17, 21], Lorenz chaotic system [18, 19, 21] and Chen chaotic system [20, 21] has been explored. It may be noted though the dimension of these systems are increased they reflect chaotic, periodic, quasi-periodic behaviours but do not reflect hyperchaotic behaviour.

Many researches have been done for the electronic implementation of chaotic and hyperchaotic systems also. In [22, 23, 24, 25, 26, 27], the chaotic or hyperchaotic system are implemented in NI Multisim software whereas the NI LabVIEW software is used in [26, 28, 29, 30, 31]. In [32], the implementaion is done using Matlab-Simulink and Orcad-PSpice.

For nearly three decades chaotic and hyperchaotic systems have received increasing attention in the application of secure communication. There have been so many techniques of synchronisation and encryption introduced to use chaotic and hyperchaotic systems to enforce safe communication. Due to the exponential divergence of the nearby trajectories of chaotic systems, it might seem surprising to have two chaotic systems evolving in synchrony. However, the synchronisation of coupled or driven chaotic oscillators is a phenomenon that has been well established experimentally and theoretically reasonably well understood. synchronisation can pose a number of forms depending on the existence of the interacting systems, the type of coupling, and the proximity of the systems [33].

In 1990, Pecora and Carroll produced the idea for chaos synchronisation [3]. Many types of synchronisation phenomenon have been studied afterwards. In [34], by using adaptive controller, synchronisation between transmitter and receiver is maintained and message signal using chaotic parameter modulation is reported [35, 36, 37, 38]. Although chaotic switching is more secured against noise than chaotic masking, it can have a lower information transmission rate than the methods which transmit analog signals directly because of the receiver has to wait till synchronisation achieved before the next bit is transmitted [39, 40, 41, 42, 43]. The above motivations enforce the following objectives:

1. To design a new chaotic system with real state variables.

2. To analyse a new chaotic system with complex state variables to that of real state variables.

In the next Section, a third order chaotic system is proposed and its different properties are analyzed and discussed.

3. DESCRIPTION AND ANALYSIS OF 3D CHAOTIC SYSTEM

The dynamics of a novel three dimensional autonomous chaotic system is given as:

$$\begin{cases} \dot{u} = av - bu \\ \dot{v} = uw + av \\ \dot{w} = -dw - cuv \end{cases} \quad (2)$$

where $u(t)$, $v(t)$ and $w(t)$ are real state variables and a, b, c and d are real positive parameters. The system (2) has two nonlinear terms and is governed by four parameters. When $a = 2$, $b = 3.5$, $c = 3$ and $d = 1$, system (2) shows typical double wing chaotic attractors and time series behaviours of the proposed system are shown in Fig. 7 which confirm the chaotic behaviour. The initial condition of proposed system (2) is considered as $[u(0)\ v(0)\ w(0)]^T = [-0.1\ -0.1\ 0.1]^T$ for simulation. The proposed chaotic system shows double-wing butterfly chaotic attractor.

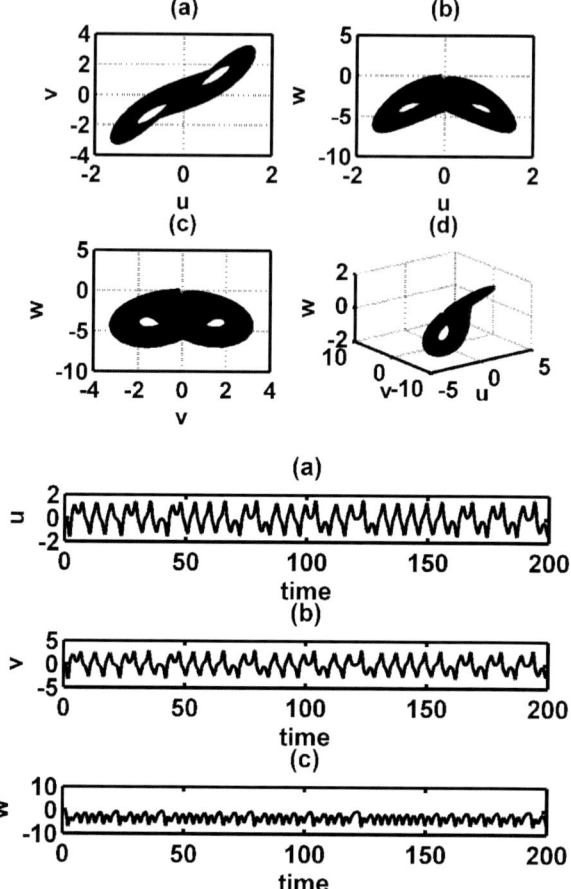

Figure 7. Phase plane and time series behaviours of real valued chaotic system (2).

3.1. Equilibrium Points

The equilibria of proposed chaotic system (2) is found by solving $\dot{u} = 0$, $\dot{v} = 0$ and $\dot{w} = 0$. Three equilibrium points of proposed system (2) are obtained as $E_1 = (0, 0, 0)$, $E_2 = (\sqrt{\frac{ad}{c}}, b\sqrt{\frac{d}{ac}}, -b)$ and $E_3 = (-\sqrt{\frac{ad}{c}}, -b\sqrt{\frac{d}{ac}}, -b)$. The generalised Jacobian matrix of system (2) to study the stability of equilibrium points is calculated as:

$$J_E = \begin{bmatrix} -b & a & 0 \\ w & a & u \\ -cv & -cu & -d \end{bmatrix}$$

The characteristic equation about equilbrium point E_1 is given by:

$$(\lambda + b)(\lambda + d)(a - \lambda) = 0 \qquad (3)$$

The eigenvalues about E_1 are calculated as $\lambda_1 = -3.5$, $\lambda_2 = 2$ and $\lambda_3 = -1$, hence E_1 is an unstable fixed point. Similarly, the eigenvalues about E_2 are calculated as $\lambda_1 = -2.9337$, $\lambda_2 = 0.2168 - j2.1737$ and $\lambda_3 = 0.2168 + j2.1737$ and reflect saddle-focus nature of equilibrium point E_2. Since E_2 and E_3 equilibria are symmetric about w axis, then nature of equilibrium point E_3 is also saddle-focus.

3.2. Symmetry and Invariance

The proposed system (2) is symmetric about w axis since it is invariant under the transformation $(u, v, w) \rightarrow (-u, -v, w)$. Hence, if (u, v, w) is a solution of the system (2), then $(-u, -v, w)$ is also the solution of the proposed system.

3.3. Dissipation

The divergence of the proposed system (2) is calculated as:

$$\nabla V = \frac{\partial \dot{u}}{\partial u} + \frac{\partial \dot{v}}{\partial v} + \frac{\partial \dot{w}}{\partial w} = a - b - d \qquad (4)$$

If $(a - b - d) < 0$, then the proposed chaotic system (2) is dissipative. For parameter values, $a = 2$, $b = 3.5$, $c = 3$ and $d = 1$, the divergence, $\nabla V = -2.5$ shows the proposed system (2) is a dissipative system.

3.4. Lyapunov Exponents

Lyapunov exponent measures the exponential rates of divergence and convergence of neighboring trajectories in phase space of the dynamical system. Lyapunov spectrum of proposed system (2) with d as a varying parameter is shown in Fig. 8. The parameter d varies between the range

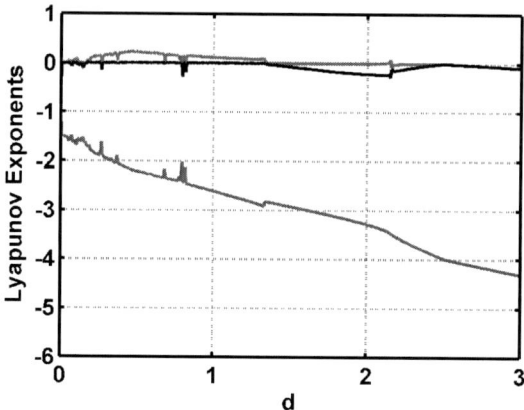

Figure 8. Lyapunov spectrum of system (2) where parameter d varies from 0 to 3.

$0 \leq d \leq 3$. For parameter values $a = 2$, $b = 3.5$, $c = 3$ and $d = 1$, Lyapunov exponents are obtained as $L_1 = 0.115$, $L_2 = 0$ and $L_3 = -2.615$. The system (2) exhibits chaotic, periodic and stable behaviours for different range of d parameter. The system shows chaotic behaviour for $d \in [0.19, 1.33]$, periodic behaviour for the range of $d \in (1.33, 2.165]$ and stable nature persists for rest of the d value.

3.5. Lyapunov Dimension

Lyapunov dimension relates dynamics of the system to its attractor geometry. Kaplan and Yorke [43] proposed a dimension based on Lyapunov exponents and known as Kaplan-Yorke dimension or Lyapunov dimension D_L and is defined as:

$$D_L = k + \frac{1}{|L_{k+1}|} \sum_{i=1}^{k} L_i \qquad (5)$$

where L represents Lyapunov exponents and k is the largest integer for which sum of the first k exponents is positive. Using (5), Lyapunov dimension of proposed chaotic system (2) is calculated as:

$$D_L = 2 + \frac{0.115}{2.615} = 2.044 \qquad (6)$$

$D_L = 2.044$ is the estimate of dimension of the volume that neither grows nor decays.

4. Design of Chaotic System with Complex State Variables

Let u, v and w states of the chaotic system (2) be complex valued and defined as:
$$\begin{cases} u = \omega_1 + i\omega_2 \\ v = \omega_3 + i\omega_4 \\ w = \omega_5 + i\omega_6 \end{cases} \quad (7)$$

where a, b, c and d are real positive constants and $i = \sqrt{-1}$. Using (7), complex value of u, v and w in terms of ω_1, ω_2, ω_3, ω_4, ω_5, ω_6, the dynamics of system (2) is modified as:

$$\begin{cases} \dot{\omega}_1 + i\dot{\omega}_2 = a(\omega_3 + i\omega_4) - b(\omega_1 + ib\omega_2) \\ \dot{\omega}_3 + i\dot{\omega}_4 = (\omega_1 + i\omega_2)(\omega_5 + i\omega_6) + a(\omega_3 + i\omega_4) \\ \dot{\omega}_1 + i\dot{\omega}_4 = -d(\omega_5 + i\omega_6) - c(\omega_1 + i\omega_2)(\omega_3 + i\omega_4) \end{cases} \quad (8)$$

A six dimensional dynamics in terms of real valued is obtained in (9) by comparing the real and imaginary parts in the left and right hand sides of (8).

$$\begin{cases} \dot{\omega}_1 = a\omega_3 - b\omega_1 \\ \dot{\omega}_2 = a\omega_4 - b\omega_2 \\ \dot{\omega}_3 = \omega_1\omega_5 - \omega_2\omega_6 + a\omega_3 \\ \dot{\omega}_4 = \omega_1\omega_6 + \omega_2\omega_5 + a\omega_4 \\ \dot{\omega}_5 = -d\omega_5 - c\omega_1\omega_3 + c\omega_2\omega_4 \\ \dot{\omega}_6 = -d\omega_6 - c\omega_1\omega_4 - c\omega_2\omega_3 \end{cases} \quad (9)$$

where ω_1, ω_2, ω_3, ω_4, ω_5, ω_6, are state variables. a, b, c and d are real and positive parameters. The analysis of the system (9) is discussed in the next section.

5. ANALYSIS OF COMPLEX STATE VARIABLES BASED SYSTEM

The time series behaviour of the complex valued chaotic system (9) is shown in Fig. 9. The phase plane behaviours of the complex valued chaotic system is shown in Fig. 10. Figures 9 and 10 confirms the chaotic behaviour of system (4).

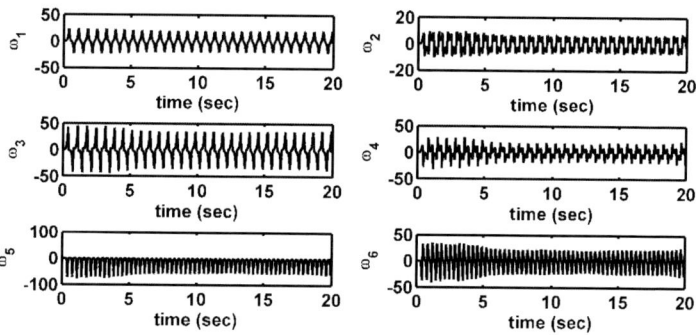

Figure 9. Time series plot of state variables of system (9) for parameter values $a = 15.6$, $b = 19$, $c = 3$ and $d = 10$.

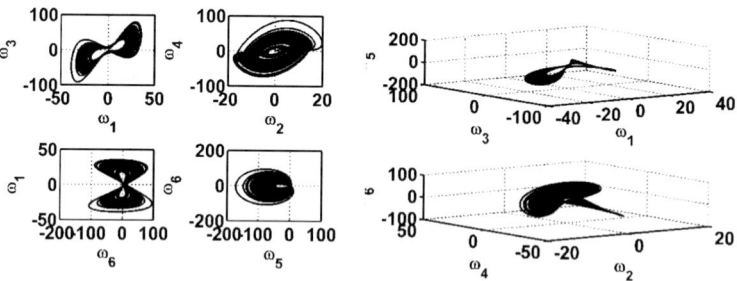

Figure 10. Phase plane behaviour of system (9) for parameter values $a = 15.6$, $b = 19$, $c = 3$ and $d = 10$.

Different properties of complex state variables based chaotic system (9) are analysed and discussed further.

5.1. Equilibria

The equilibrium points of complex state variables based system (8) is calculated by solving (10).

$$\begin{cases} a\omega_3 - b\omega_1 = 0 \\ a\omega_4 - b\omega_2 = 0 \\ \omega_1\omega_5 - \omega_2\omega_6 + a\omega_3 = 0 \\ \omega_1\omega_6 + \omega_2\omega_5 + a\omega_4 = 0 \\ -d\omega_5 - c\omega_1\omega_3 + c\omega_2\omega_4 = 0 \\ -d\omega_6 - c\omega_1\omega_4 - c\omega_2\omega_3 = 0 \end{cases} \quad (10)$$

Generalised equilibria ω_{e_0}, ω_{e_+} and ω_{e_-} at $\omega_2 = 0$, $i\sqrt{\dfrac{ad}{c}}$, $-i\sqrt{\dfrac{ad}{c}}$, respectively, are obtained as:

$$\omega_{e_0} = \left[\omega_1,\ 0,\ \frac{b}{a}\omega_1,\ 0,\ -\frac{cb}{ad}\omega_1^2,\ 0\right]$$

$$\omega_{e_+} = \left[\omega_1,\ ik,\ \frac{b}{a}\omega_1,\ \frac{ib}{a}k,\ -b - \frac{cb}{ad}\omega_1^2,\ -i2b\sqrt{\frac{c}{ad}}\omega_1\right] \quad (11)$$

$$\omega_{e_-} = \left[\omega_1,\ -ik,\ \frac{b}{a}\omega_1,\ \frac{-ib}{a}k,\ -b - \frac{cb}{ad}\omega_1^2,\ i2b\sqrt{\frac{c}{ad}}\omega_1\right]$$

where $\omega_1 = 0, \pm\sqrt{3\omega_2^2 + \dfrac{ad}{c}}$ and $k = \sqrt{\dfrac{ad}{c}}$.

When $\omega_1 = 0$, the equilibria of derived system (10) are obtained as:

$$\begin{cases} \omega_{e_1} = [0,\ 0,\ 0,\ 0,\ 0,\ 0] \\ \omega_{e_2} = \left[0,\ ik,\ 0,\ i\dfrac{b}{a}k,\ -b,\ 0\right] \\ \omega_{e_3} = \left[0,\ -ik,\ 0,\ -i\dfrac{b}{a}k,\ b,\ 0\right] \end{cases}$$

Similarly, when $\omega_1 = \pm\sqrt{3\omega_2^2 + \dfrac{ad}{c}}$, we have obtained three-three equilibria for each positive and negative values of ω_1, respectively, and given

as:
$$\begin{cases} \omega_{e_4} = \left[k,\ 0,\ \dfrac{b}{a}k,\ 0,\ -b,\ 0\right] \\ \omega_{e_5} = \left[i\sqrt{2}k,\ ik,\ i\dfrac{b}{a}\sqrt{2}k,\ i\dfrac{b}{a}k,\ b,\ 2\sqrt{2}b\right] \\ \omega_{e_6} = \left[i\sqrt{2}k,\ -ik,\ i\dfrac{b}{a}\sqrt{2}k,\ -i\dfrac{b}{a}k,\ b,\ 2\sqrt{2}b\right] \end{cases}$$

$$\begin{cases} \omega_{e_7} = \left[-k,\ 0,\ -\dfrac{b}{a}k,\ 0,\ -b,\ 0\right] \\ \omega_{e_8} = \left[-i2k,\ ik,\ -i\dfrac{2b}{a}k,\ i\dfrac{b}{a}k,\ b,\ 2\sqrt{2}b\right] \\ \omega_{e_9} = \left[-i2k,\ -ik,\ -i\dfrac{2b}{a}k,\ -i\dfrac{b}{a}k,\ b,\ 2\sqrt{2}b\right] \end{cases}$$

where $k = \sqrt{\dfrac{ad}{c}}$.

The gerneralised Jacobian matrix of system (9) is given as:

$$J_E = \begin{bmatrix} -b & 0 & a & 0 & 0 & 0 \\ 0 & -b & 0 & a & 0 & 0 \\ \omega_5 & -\omega_6 & a & 0 & \omega_1 & -\omega_2 \\ \omega_6 & \omega_5 & 0 & a & \omega_2 & \omega_1 \\ -c\omega_3 & c\omega_4 & -c\omega_1 & c\omega_2 & -d & 0 \\ -c\omega_4 & c\omega_3 & -c\omega_2 & -c\omega_1 & 0 & -d \end{bmatrix}$$

The characteristic polynomial about ω_{e_1} is given as:

$$(\lambda + b)^2 (\lambda + d)^2 (a - \lambda)^2 = 0 \quad (12)$$

The eigenvalues at equilibrium point ω_{e_1} are obtained as $\lambda_1, \lambda_2 = -19$, $\lambda_3 = \lambda_4 = 15.6$ and $\lambda_5 = \lambda_6 = -10$, hence ω_{e_1} is an unstable fixed point.

Similarly, eigenvalues at equilibria ω_{e_2}, ω_{e_3}, ω_{e_4}, ω_{e_5}, ω_{e_6} are calculated and it is observed that the behaviour about ω_{e_2}, ω_{e_3}, ω_{e_4}, ω_{e_5}, ω_{e_6}, ω_{e_7}, ω_{e_8} is saddle-focus whereas the nature of equilibrium point ω_{e_9} is unstable focus.

5.2. Symmetry and Invariance

The complex state variables based system (9) is symmetric about (ω_5-ω_6) plane due to invariance of system dynamics under (ω_1, ω_2, ω_3, ω_4, ω_5,

$\omega_6) \to (-\omega_1, -\omega_2, -\omega_3, -\omega_4, \omega_5, \omega_6)$. Hence, if $(\omega_1, \omega_2, \omega_3, \omega_4, \omega_5, \omega_6)$ is the solution of system, then $(-\omega_1, -\omega_2, -\omega_3, -\omega_4, \omega_5, \omega_6)$ is also the solution of the system (9).

5.3. Dissipation

The divergence of complex valued system (9) is calculated as:

$$\nabla V = \sum_{i=1}^{6} \frac{\partial \dot{\omega}_i}{\partial \omega_i} = 2(a - b - d) \qquad (13)$$

If $(a - b - d) < 0$, then complex valued system (9) is dissipative. For parameters $a = 15.6$, $b = 19$, $c = 3$ and $d = 10$, divergence $\nabla V = -26.8$ confirms the dissipating nature of system (9), so as to system (8).

5.4. Lyapunov Exponent and Dimension

Lyapunov spectrum is plotted by varying one parameter at a time and other parameters are kept fixed. The Lyapunov exponents of complex valued systems (9) for d as a varying parameter is shown in Fig. 11. Lyapunov spectrum is plotted by varying d parameter between $3.05 \leq d \leq 15$. The system shows hyperchaotic behaviour in the range of $d \in (5.96, 11.55)$, the chaotic behaviour for the range of $d \in (3.575, 3.7)$, and quasi-periodic behaviour for $d \in [3.05, 3.575) \cup (3.7, 5.96) \cup (11.55, 15]$. Lyapunov exponents of the system at parameter values, $a = 15.6$, $b = 19$, $c = 3$ and $d = 10$ are calculated as $L_6 = 0.3195$, $L_5 = 0.3214$, $L_4 = 0.0015 \approx 0$, $L_2 = 0.0002 \approx 0$, $L_1 = 0.0003 \approx 0$ and $L_3 = -5.2236$. The system (9) has two positive Lyapunov exponents (L_5 & L_6) along with three zero Lyapunov exponents (L_4, L_2, & L_1) reveals hyperchaotic 3-torus nature.

Lyapunov dimension of complex valued chaotic system (9) is calculated using (5) and given as:

$$D_L = 5 + \frac{0.64298}{5.2236} = 5.1248. \qquad (14)$$

The advantages of complex state variables based system over real state variables based system are compared and summarised in Table 1.

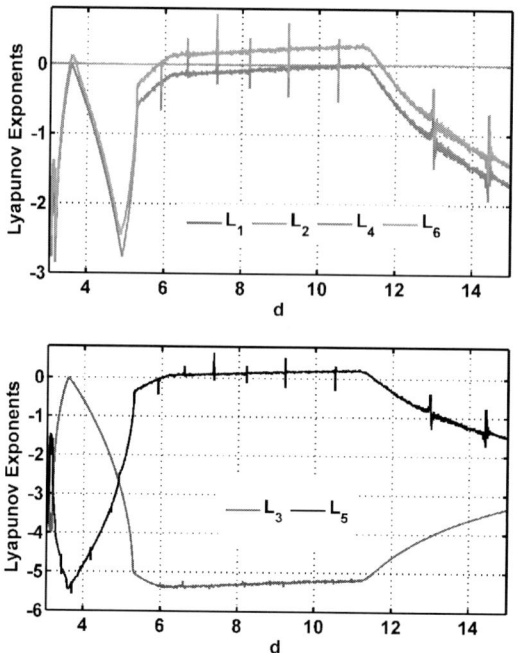

Figure 11. Lyapunov spectrum plot for the system (9) where parameter d varies from 3.05 to 15.

Table 1. Comparison between the proposed 3D real valued chaotic system and derived 6D complex valued hyperchaotic system

Comparison Parameters	3D chaotic system	6D hyperchaotic system
No. of state variables	3	6
No. of Equilibria	3	9
Largest Lyapunov Exponent	0.1147	0.2923
Lyapunov Dimension	2.4396	5.1248
Magnitude range of state variables	$\in [-5, +10]$	$\in [-200, +100]$
Behaviours	Chaotic, Periodic and Stable	Hyperchaotic 3-torus, Chaotic, Quasiperiodic and Stable

CONCLUSION

A novel three dimensional chaotic system is analysed with real and complex state variables. A six dimensional hyperchaotic system is derived by using complex value of state variables in the proposed three dimensional system. The complex valued system has eight nonlinear terms, four parameters and nine equilibria. The six dimensional system (Complex valued system) shows hyperchaotic 3-torus behaviour which is very rare in the literature as per our best knowledge. The largest Lyapunov exponent and Lyapunov dimension of the complex valued system is 2.5 and 2.0 times, respectively, greater than to its real valued system (please refer Table 1). Moreover, it may be noted that the complex valued system also retained chaotic and periodic behaviours. The magnitude range of state variables for the complex valued system is 20 times more than the real valued chaotic system. Therefore, the derived complex valued hyperchaotic system can be used for the encryption of massage signal in communication and explored in Chapter 4.

REFERENCES

[1] Yurkon G. T., "*Introduction to chaos and it's real world applications*", Cleveland State University, May 28, 1997.

[2] Saha P., Banerjee S. and Chowdhury A. R., "Chaos, signal communication and parameter estimation", *Physics Letters*, 326:2, 133-139, (2004).

[3] Pecora L. M. and Carrol T. L., "Synchronisation in chaotic system", *Physical Review Letters*, 64:8, 821-825, (1990).

[4] Rossler O. E., "An equation for hyperchaos," *Physics Letters A*, 71:2, 155-157, (1979).

[5] Ozer A. B. and Akin E., "Tools for detecting Chaos", *Sakarya University Journal of Science*, 9:1, 60-66, (2005).

[6] Kumar R. R., Sampath A. and Indumathi P., "Improving security of communication systems using chaos", *Indian Journal of Science and Technology*, 4:5, 561-565, (2011).

[7] Rauf K. and Yasir M., "Chaos based optical communication", *International Journal of Computer and Communication Engineering*, 2:2, 97-100, (2013).

[8] Cenys A., Tamasevicius A. and Lindberg E., "Hyperchaos in coupled Colpitts oscillators," *Chaos, Solitons and Fractals*, 17:2, 349-353, (2003).

[9] Li Y., Tang W. and Chen G., "Generating hyperchaos via state feedback control," *International Journal of Bifurcation and Chaos*, 15:10, 3367-3375, (2005).

[10] Wang G., Zhang X., Zheng Y. and Li Y., "A new modified hyperchaotic Lu system," *Physica A: Statistical Mechanics and Its Applications*, 371:2, 260-272, (2006).

[11] Barboza R., "Dynamics of a hyperchaotic Lorenz system," *International Journal of Bifurcation and Chaos*, 17:12, 4285-4294, (2007).

[12] Wang J. and Chen Z., "A novel hyperchaotic system and its complex dynamics," *International Journal of Bifurcation and Chaos*, 18, 3309-3324, (2008).

[13] Vaidyanathan S., Pham V.T. and Volos C.K., "A 5-D hyperchaotic Rikitake dynamo system with hidden attractors," *European Physical Journal Special Topics*, 224:8, 1575-1592, (2015).

[14] Liu Y. J., Yang Q.G. and Pang G.P., "A hyperchaotic system from the Rabinovich system," *Journal of Computer Applied Mathematics*, 234, 101-113, (2010).

[15] Thomas R., Basios V., and Rossler O. E., "Hyperchaos of arbitrary order generated by a single feedback circuit and the emergence of chaotic walks," *Chaos: An Interdisciplinary Journal of Nonlinear Science*, 14:3, 669-674, (2004).

[16] Mahmoud E. E. and Abood F. S., "A new nonlinear chaotic complex model and its complex antilag synchronisation," *Complexity*, 2017, 1-13, (2017).

[17] Yadav V. K., Kumar R., Leung A. Y. T. and Das S., "Dual phase and dual anti-phase synchronisation of fractional order chaotic systems in real and complex variables with uncertainties," *Chinese Journal of Physics*, 57, 282-308, (2019).

[18] Sun B., Li M. and Liu J., "The characteristics and self-time-delay synchronisation of two-time-delay complex Lorenz system," *Journal of the Franklin Institute*, 356, 334-350, (2019).

[19] Moghtadaei M. and Golpayegani M. R. H., "Complex dynamic behaviors of the complex Lorenz system," Scientia Iranica, 19, 733-738, (2012).

[20] Yadav V. K., Prasad G., Srivastava M. and Das S., "Combination-combination phase synchronisation among non-identical fractional order complex chaotic systems via nonlinear control," *International Journal of Dynamics and Control*, 7, 330-340, (2019).

[21] Mahmoud E. E. and Adwani M. A., "Dynamical behaviors, control and synchronisation of a new chaotic model with complex variables and cubic nonlinear terms," *Results in Physics*, 7, 1346-1356, (2017).

[22] Lassoued A. and Boubaker O., "Dynamic analysis and circuit design of a novel hyperchaotic system with fractional-order terms," *Complexity*, 2017, 1-10, (2017).

[23] Ibrahim K. M., Jamal R. K. and Ali F. H., "Chaotic behaviour of the Rossler model and its analysis by using bifurcations of limit cycles and chaotic attractors," *Journal of Physics*, 1003, 012099-012106, (2018).

[24] Hammouch Z. and Mekkaoui T., "Circuit design and simulation for the fractional-order chaotic behavior in a new dynamical system," *Complex and Intelligent Systems*, 4, 251-260, (2018).

[25] Xiong L., Lu Y. J. and Gupta P., "Design and hardware implementation of a new chaotic secure communication technique," PLosONE, 11:8, 1-19, (2016).

[26] Bulut G. G., Sahin M. E. and Guler H., "An implementation of chaotic circuits with Multisim-LabVIEW," *International Advanced Researches and Engineering Journal*, 2:3, 304-308, (2018).

[27] Singh P. P., Singh J. P. and Roy B. K., "Implementation for synchronisation of Chen and Liu-Yang chaotic systems using SMC and active control schemes," *International Journal of Control Theory and Applications*, 9, 159-169, (2017).

[28] Bernardara N., Lafaurie A. and Saad F. I., "On the Rucklidge time-delayed chaotic system for nonlinear double convection: Adaptive control, synchronisation and LabVIEW implementation," *Annals of Electrical and Electronic Engineering*, 2:6, 10-19, (2019).

[29] Gupta J., Kosta S. P. and Mor P., "Simulation of non-autonomous chaotic circuit on LabVIEW using nonlinear electrolytic device," *International Journal of Electronics Engineering*, 3:2, 215-220, (2011).

[30] Yang H., "A new chaotic attractor and its digital implementation based on LabVIEW," *Applied Mechanics and Materials*, 278, 54-57, (2013).

[31] Elkholy M. M., Hennawy H. M. E. and Elkouny A., "Simulation and digital implementation of eight dimensional hyper chaotic system for secured chaotic communication," *International Journal of Engineering Research & Technology*, 4:3, 482-486, (2015).

[32] Alsafasfeh Q. H. and Arni M. S., "A new chaotic behavior from Lorenz and Rossler systems and its electronic circuit implementation," *Circuits and Systems*, 2, 101-105, (2011).

[33] Slotine J. E., and Li W., *Applied Nonlinear Control*, Prentice Hall Inc., New Jersey (1991).

[34] Yang T. and Chua L., "Secure Communication via chaotic parameter modulation", *IEEE Transactions on circuits and systems: Fundamental Theory and applications*, 43:9, 817-819, (1996).

[35] Angeli A. D., Genesio R. and Tesi A., "Dead-beat chaos synchronisation in discrete-time systems", *IEEE Transactions on circuits and systems: Fundamental Theory and applications*, 42:1, 54-56, (1995).

[36] Chen S. and Lu J., "Synchronisation of an uncertain unified chaotic system via adaptive control", *Chaos, Solitons and Fractals*, 14:4, 643-647, (2002).

[37] Xue C., Jiang N., Lv Y. and Qiu Kun, "Secure key distribution based on dynamic chaos synchronisation of cascaded semiconductor laser systems", *IEEE Transactions on Communications*, 65, 312-319, (2017).

[38] Li C., Sprott J. C., Akgul A. and Zhao Yibo, "A new chaotic oscillator with free control", *Chaos: An Interdisciplinary Journal of Nonlinear Science*, 27, 083101-083106, (2017).

[39] Wen G., Chen M. Z. Q. and Yu X., "Event-triggered master-slave synchronisation with sampled-data communication", *IEEE Transactions on Circuits and Systems*, 63:3, 304-308, (2016).

[40] Souaia M. A., Trabelsi H. and Saad K. B., "Synchronisation of the Liu chaotic system and its application in secure communication", *2017 International Conference on Control, Automation and Diagnosis (ICCAD)*, Hammamet, 17, 434-438, (2017).

[41] Aromataris G. and Lodi V. A., "Assessing Security of Chaos Communications Against Eavesdropping by Pump Detection", *IEEE Journal of Quantum Mechanics*, 55:2, 101-104, (2019).

[42] Singh P. P., Singh J. P., Borah M. and Roy B. K., "On the construction of a new chaotic system", *IFAC-PapersOnLine*, 49:1, 522-525, (2016).

[43] Singh P. P. and Roy B. K., "Memristor based novel complex chaotic system and its projective synchronisation using nonlinear active control technique", *European Physical Journal Special Topics*, 228, 2197-2214, (2019).

In: Nonlinear Systems
Editor: P. P. Singh
ISBN: 978-1-68507-660-3
© 2022 Nova Science Publishers, Inc.

Chapter 2

PROJECTIVE SYNCHRONISATION OF COMPLEX VALUED HYPERCHAOTIC SYSTEMS AND ENCRYPTION TECHNIQUES IN SECURE COMMUNICATIONS

Sugandha[*] *and Piyush Pratap Singh*[†]
Department of Electrical Engineering,
NIT Meghalaya, Shillong, Meghalaya, India

Abstract

In this chapter, firstly, projective synchronisation between two identical complex state variables based hyperchaotic systems using nonlinear active control technique is discussed and required stability condition is derived using Lyapunov stability theory. Secondly, two new encryption, Diffie-Hellman key exchange-masking and modulation-masking, techniques are proposed and analysed in the application of secure communication. These two encryption techniques are obtained by cascading of modulation technique with Diffie-Hellman key exchange and modulation techniques and are demonstrated with complex state variables based hyperchaotic system in secure communication as an illustrative example. Finally, the traditional masking, proposed D-H key exchange-masking and

---
[*]Corresponding Author's E-mail: itz.sugandha11@gmail.com.
[†]Corresponding Author's E-mail: piyushpratapsingh@gmail.com.

mod-mask encryption techniques are compared where both the encryption techniques are performing better. Simulation is presented in MATLAB environment to demonstrate the effective verification of the theoretical approach. Simulation results oblige that the objectives are achieved successfully.

1. INTRODUCTION

In 1990, Pecora and Carroll produced the idea for chaos synchronisation [1]. Synchronisation in nonlinear dynamic systems is a well-known phenomenon that has been intensively studied over the past few years due to all its possible application in data communication [2, 3]. The process of synchronisation is now widely used for encoding and decoding messages. While communicating messages or signal, security is one of the important matter concerned. Hence, the encoding and decoding method should be arranged so that no interception is possible [4]. Two chaotic systems, transmitter and receiver need to be synchronized for the communication. Chaos based synchronisation between the transmitter and receiver plays an important role in secure communication [5].

1.1. Synchronisation of Chaotic System

Types of chaos synchronisation is classified as follows [5]:

1. *Identical synchronisation*: Identical synchronisation is a type of synchronisation when two identical chaotic systems are coupled mutually or either of the systems is driven by the other one.

2. *Generalized synchronisation*: In generalized synchronisation, the dynamic variables which define the state of the systems are affiliated with each other using a transform function so that they are identical to each other after transformation.

3. *Phase synchronisation*: Phase synchronisation is the case when two non-identical systems are coupled mutually with respect to phase whereas the amplitudes remain unsynchronized.

4. *Anticipated and lag synchronisation*: Anticipated and lag synchronisation is a form of synchronisation where a time difference described

by delay differential equations, relates the synchronisation state of the systems to one another. In that form, one oscillator 's dynamics follow the other one's dynamics.

5. *Amplitude envelop synchronisation*: Amplitude envelope synchronisation is a soft form of synchronisation in which there is no synchronisation between phase and amplitude, but the systems present a periodic envelope with the same frequency in both systems.

1.2. Chaotic Communication Schemes

Today, communication security is an important issue. In the framework of drive-response, one chaotic system can be considered as a drive system, and another chaotic system is considered as a response system. The dynamical behaviours of drive and response systems can be synchronize when the response system is driven by a control input. By using this drive-response based synchronisation framework, various chaotic systems are synchronised and applied to secure communication systems.

In chaos based communication, the data is transmitted in aforementioned three schemes of data transmission [6]:

1. *Chaotic masking*: In chaotic masking, message signal is added to the chaotic drive signal and transmitted over the channel.

2. *Chaotic modulation*: Chaotic modulation message signal is multiplied with the chaotic carrier.

3. *Chaos switching*: In chaos switching, the chaotic signals are generated for all symbols in the alphabet set similar to the shift keying methods.

2. LITERATURE REVIEW AND MOTIVATION

In 1979, Otto Eberhard Rossler had discovered hyperchaos phenomenon with a four dimensional autonomous system and known as hyperchaotic Rossler system [7]. Other hyperchaotic systems such as Colpitt's, Chen's, Lu's, Lorenz's were also constructed via. anti-control in [8, 9, 10, 11, 12, 13, 14, 15]. Design of hyperchaotic system by using complex state variables in place of real state variable and its synchronisation has been explored by researchers [16, 17, 18, 19, 20, 21, 22, 23]. Many types of synchronisation phenomenon have been studied after the remarkable work by Pecora and Carroll. In [24], synchronisation between transmitter and receiver is maintained by using adaptive controller and message signal using chaotic parameter modulation is reported. Although chaotic switching is more secured against noise than chaotic masking, it can have a lower information transmission rate than the methods which transmit analog signals directly. This is because the receiver has to wait until synchronisation and de-synchronisation have been achieved before the next bit is transmitted.

The dead-beat synchronisation or exact synchronisation in finite time is investigated to synchronize discrete-time chaotic systems [25]. Systems that achieve synchronisation in a finite number of steps are called as dead-beat synchronizing. Ddead-beat synchronizing systems is always immediate. Due to the deterministic nature of chaotic motions, in fact, once dead-beat synchronisation has been achieved the two systems will remain synchronized regardless of the presence of the synchronisation signal. But, the drawback of this technique is that the synchronisation string has to be repeated at the beginning of every cycle.

The design process for chaos synchronisation for uncertain systems using adaptive method controller is reported in [26]. The adaptive synchronisation technique is used for uncertain chaotic system which makes the technique robust to parameter variations. Chaos synchronisation in two identical cascaded semiconductor laser systems (CSLSs) subject to common light injection and random phase modulated optical feedback is proposed in [27].

A chaotic oscillator with free and unlimited control of its frequency, amplitude, and polarity by means of simple adjustable resistors and an adjustable dc source is proposed in [28]. The frequency and amplitude control is achieved by two independent potentiometers and the polarity control is achieved by adjustable dc source. synchronisation of heterogeneous master-slave coupling systems consisting of a high-dimensional master system but a low-dimensional slave system is proposed in [29]. Only the master system's sampled-data outputs are used to design observer and controller. The present method of design is therefore more favorable than those depending on the master system's sampled-data states.

Synchronisation of two chaotic systems in the drive-response configuration is proposed in [30]. This means that two chaotic systems are coupled in such a way that the response system trajectories converge, in finite time, to those of the drive system in which only one controller is required and only one signal is required to be sent to the controller from the drive system.

A method is evaluated that could be tried by an eavesdropper to force the standard two-laser scheme of chaos private transmission in [31]. This approach is based on detecting the voltage emanating from a semiconductor laser junction and uses two lasers: one is synchronized with the transmitter (thus, reproducing chaos and rejecting the message), the other is chosen and trimmed to copy the transmitter (i.e., reproducing both chaos and message). The message is then obtained by cancellation of the difference and disorder. The eavesdropper scheme is based on the detection of pumps. Bhalekar-Gejji (BG) chaotic system is chaotified in order to generate chaos in [32]. Chaos control or anti-control refers to purposefully manipulating chaotic dynamical behaviours of a complex nonlinear system.To drive the stable BG system into chaotic mode, a simple linear partial state feedback controller is developed. The proposed new chaotic system has more complex behaviour than the BG chaotic systems.

A concept of magnifying glass is introduced in [34] which is used to enlarge and observe minor parameter mismatch so as to increase the sensitivity of the system.The mismatch parameter is a large number, we can see that multiple times the signal is enlarged, which implied that the mismatch parameters could be enlarged. Therefore even a small discrepancy in the parameters can result in a significant decryption error, resulting in a decryption key sequence which is not the same as the key signal for encryption.The slave system is modelled by an unknown input observer, in which the transmitted information is the unknown input. As in the general observer-based synchronisation paradigm, the information is recovered if the master and slave systems robustly synchronize. In the context of unknown-input observers, this is important to estimating the master's states and the unknown inputs.

Many encryption techniques have also been proposed using chaotic or hyperchaotic signals as a carrier to hide the data. One of the earliest technique, called additive masking was based on simply adding the message to one of the chaotic states of the transmitter and extracting the message by simply subtracting it from the synchronised state at the receiving end. However, the technique is found out to be successfully to hide the message but was quite impractical because of its sensitivity to channel noise and parameters mismatch between both the transmitter and the receiver. In addition, this method proved to have poor security [36].

Many modulation methods also have been proposed such as Chaos Shift Keying (CSK), Differential CSK (DCSK), Chaos On Off Keying (COOK), FM-DCSK, Delay-CSK, and so on [37]. Most of these techniques are classified into coherent and noncoherent types. In coherent type, an identical chaotic carrier has to be designed at the reciever using synchronisation technique. In non coherent type technique, the exact replica of chaotic carrier need not to be reproduced but instead the data is recovered in binary form using bit-energy estimation. This type is more reliable but found to be easily detected by unauthorised receivers [38].

Furthermore, the aforementioned methods are mostly useful for digital communication except chaotic masking and chaotic parameter modulation [39]. But, there are many breaking of masking and modulation have also been reported [40, 41, 42, 43, 44, 45]. In order to overcome this, more reliable and robust encryption techniques need to be explored in the field of analog communication. Despite several years of study of this phenomenon and its applications, there are still many fundamental issues open. The above mentioned issues motivated to work upon these for further improvement.

Literature reveals that the field of communication is always in need of alternate ways of producing different kind of signals, variation in the scheme of synchronisation and encryption techniques to establish secure communication. The field of chaos is incorporated to achieve the aforementioned objectives but there are many breaking of masking and modulation have also been reported [40, 41, 42, 43, 45, 46]. In order to overcome this, more reliable and robust encryption techniques need to be explored in the field of analog communication. In the direction for improvement of communication, several design on new chaotic systems has been reported in the literature and they don't have sufficient complex behaviour as required in the communication.The chaotic system with complex state variables is exlpored because the use of imaginary variables expands the possibility for the exploration of rich chaotic property that is left out with the use of real state variables.

Moreover, the properties of chaotic system like aperiodic longterm behaviour, sensitivity to initial conditions and noise-like characteristics, enforce the need of appropriate synchronisation technique to synchronise and control in the various applications. Various chaotic systems and their synchronisation and encryption techniques have been introduced but some forced techniques are still required as per desired criteria, applications and needed to be explored further.

The above motivations enforce the following objectives:

1. To achieve projective synchronisation between complex state variables based chaotic system using a modified control technique.

2. To develop a hybrid encryption techniques (D-H key exchange-masking and mod-mask) to increase the strength of the transmitted signal to prevent unethical hacking.

3. PROJECTIVE SYNCHRONISATION OF COMPLEX STATE VARIABLES-BASED HYPERCHAOTIC SYSTEMS

A dynamics of six dimensional system is rewritten in (1).

$$\begin{cases} \dot{\omega}_1 = a\omega_3 - b\omega_1 \\ \dot{\omega}_2 = a\omega_4 - b\omega_2 \\ \dot{\omega}_3 = \omega_1\omega_5 - \omega_2\omega_6 + a\omega_3 \\ \dot{\omega}_4 = \omega_1\omega_6 + \omega_2\omega_5 + a\omega_4 \\ \dot{\omega}_5 = -d\omega_5 - c\omega_1\omega_3 + c\omega_2\omega_4 \\ \dot{\omega}_6 = -d\omega_6 - c\omega_1\omega_4 - c\omega_2\omega_3 \end{cases} \quad (1)$$

where ω_1, ω_2, ω_3, ω_4, ω_5, ω_6, are state variables. a, b, c and d are real and positive parameters. The detailed analysis of the chaotic system (1) is discussed in Chapter 1. The time series and phase plane behaviours of complex valued system confirms the hyperchaotic behaviour present in the system (1).

The behaviour produced by chaotic systems/oscillators varies randomly due to its sensitivity to initial excitation and the change in control parameters. Therefore, the synchronisation between chaotic systems is essential in the field of secure communication and other chaos applications.

In each transmission, the information signal is encrypted using different encryption techniques. The synchronised chaotic systems (transmitter and receiver) enable encryption and decryption of the information signal successfully [35].

In subsection 3.1, projective synchronisation of complex state variables based hyperchaotic systems (1), one hyperchaotic system as a drive system and other as a response system, is illustrated.

3.1. Projective Synchronisation between Hyperchaotic Systems

The complex state variables based hyperchaotic system is reconsidered here as the drive system and is defined in (2):

$$\begin{cases} \dot{u}_d = av_d - bu_d \\ \dot{v}_d = u_d w_d + av_d \\ \dot{w}_d = -dw_d - cu_d v_d \end{cases} \quad (2)$$

where $u_d = \omega_{1d} + i\omega_{2d}$, $v_d = \omega_{3d} + i\omega_{4d}$ and $w_d = \omega_{5d} + i\omega_{6d}$ are the complex state variables and a, b, c and d are real constants. Subscript 'd' represents drive.

The complex state variables based hyperchaotic system with added control inputs, $(U \in R^3)$ is considered as response system and is defined as:

$$\begin{cases} \dot{u}_r = av_r - bu_r + U_1 \\ \dot{v}_r = u_r w_r + av_r + U_2 \\ \dot{w}_r = -dw_r - cu_r v_r + U_3 \end{cases} \quad (3)$$

where $u_r = \omega_{1r} + i\omega_{2r}$, $v_r = \omega_{3r} + i\omega_{4r}$ and $w_r = \omega_{5r} + i\omega_{6r}$ are the complex state variables and a, b, c and d are real constants. Subscript 'r' denotes response. The control inputs are defined as:

$$\begin{cases} U_1 = u_1 + iu_2 \\ U_2 = u_3 + iu_4 \\ U_3 = u_5 + iu_6 \end{cases} \quad (4)$$

Now, the drive (2) and response (3) systems in the form of real variables are written in equations (5) and (6), respectively.

$$\begin{cases} \dot{\omega}_{1d} = a\omega_{3d} - b\omega_{1d} \\ \dot{\omega}_{2d} = a\omega_{4d} - b\omega_{2d} \\ \dot{\omega}_{3d} = \omega_{1d}\omega_{5d} - \omega_{2d}\omega_{6d} + a\omega_{3d} \\ \dot{\omega}_{4d} = \omega_{1d}\omega_{6d} + \omega_{2d}\omega_{5d} + a\omega_{4d} \\ \dot{\omega}_{5d} = -d\omega_{5d} - c\omega_{1d}\omega_{3d} + c\omega_{2d}\omega_{4d} \\ \dot{\omega}_{6d} = -d\omega_{6d} - c\omega_{1d}\omega_{4d} - c\omega_{2d}\omega_{3d} \end{cases} \quad (5)$$

$$\begin{cases} \dot{\omega}_{1r} = a\omega_{3r} - b\omega_{1r} + u_1 \\ \dot{\omega}_{2r} = a\omega_{4r} - b\omega_{2r} + u_2 \\ \dot{\omega}_{3r} = \omega_{1r}\omega_{5r} - \omega_{2r}\omega_{6r} + a\omega_{3r} + u_3 \\ \dot{\omega}_{4r} = \omega_{1r}\omega_{6r} + \omega_{2r}\omega_{5r} + a\omega_{4r} + u_4 \\ \dot{\omega}_{5r} = -d\omega_{5r} - c\omega_{1r}\omega_{3r} + c\omega_{2r}\omega_{4r} + u_5 \\ \dot{\omega}_{6r} = -d\omega_{6r} - c\omega_{1r}\omega_{4r} - c\omega_{2r}\omega_{3r} + u_6 \end{cases} \quad (6)$$

Let the projective synchronisation error ($e \in R^3$) between the drive (5) and response (6) systems be defined as:

$$\begin{cases} e_1 = e_{\omega 1} + ie_{\omega 2} = (\omega_{1d} - k_1\omega_{1r}) + i(\omega_{2d} - k_2\omega_{2r}) \\ e_2 = e_{\omega 3} + ie_{\omega 4} = (\omega_{3d} - k_3\omega_{3r}) + i(\omega_{3d} - k_4\omega_{3r}) \\ e_3 = e_{\omega 5} + ie_{\omega 6} = (\omega_{5d} - k_5\omega_{5r}) + i(\omega_{6d} - k_6\omega_{6r}) \end{cases} \quad (7)$$

where $e_{\omega i}$ and k_i for $i = 1, 2, 3, ..., 6$ are the errors betweeen systems (5) and (6) and projective synchronisation constants respectively.
Hence, the error dynamics is written as:

$$\begin{cases} \dot{e}_{\omega 1} = ae_{\omega 3} - be_{\omega 1} - k_1 u_1 \\ \dot{e}_{\omega 2} = ae_{\omega 4} - be_{\omega 2} - k_2 u_2 \\ \dot{e}_{\omega 3} = -ae_{\omega 3} + \omega_{1d}\omega_{5d} - k_3\omega_{1r}\omega_{5r} + 2ae_{\omega 3} - \omega_{2d}\omega_{6d} + k_3\omega_{2r}\omega_{6r} - k_3 u_3 \\ \dot{e}_{\omega 4} = -ae_{\omega 4} + \omega_{1d}\omega_{6d} - k_4\omega_{1r}\omega_{6r} + 2ae_{\omega 4} - \omega_{2d}\omega_{5d} - k_4\omega_{2r}\omega_{5r} - k_4 u_4 \\ \dot{e}_{\omega 5} = -de_{\omega 5} - c[\omega_{1d}\omega_{3d} - k_5\omega_{1r}\omega_{3r} - \omega_{2d}\omega_{4d} + k_5\omega_{2r}\omega_{4r}] - k_5 u_5 \\ \dot{e}_{\omega 6} = -de_{\omega 6} - c[\omega_{1d}\omega_{4d} - k_6\omega_{1r}\omega_{4r} - \omega_{2d}\omega_{3d} - k_6\omega_{2r}\omega_{3r}] - k_6 u_6 \end{cases}$$

$$(8)$$

Theorem 1. The designed control inputs in (9) ensure the projective synchronisation between the drive (5) and response (6) hyperchaotic systems under the asymptotic stability in large or globally asymptotically stable.

$$\begin{cases} u_1 = \dfrac{1}{k_1}[ae_{\omega 3}] \\ u_2 = \dfrac{1}{k_2}[ae_{\omega 4}] \\ u_3 = \dfrac{1}{k_3}[\omega_{1d}\omega_{5d} - k_3\omega_{1r}\omega_{5r} + 2ae_{\omega 3} - \omega_{2d}\omega_{6d} + k_3\omega_{2r}\omega_{6r}] \\ u_4 = \dfrac{1}{k_4}[\omega_{1d}\omega_{6d} - k_4\omega_{1r}\omega_{6r} + 2ae_{\omega 4} - \omega_{2d}\omega_{5d} + k_4\omega_{2r}\omega_{5r}] \\ u_5 = \dfrac{-c}{k_5}[\omega_{1d}\omega_{3d} - k_5\omega_{1r}\omega_{3r} - \omega_{2d}\omega_{4d} + k_5\omega_{2r}\omega_{4r}] \\ u_6 = \dfrac{-c}{k_6}[\omega_{1d}\omega_{4d} - k_6\omega_{1r}\omega_{4r} + \omega_{2d}\omega_{3d} - k_6\omega_{2r}\omega_{3r}] \end{cases} \quad (9)$$

Proof. Let the Lyapunov function candidate is continuously differentiable function and defined as:

$$V(e_\omega) = \frac{1}{2}\sum_{i=1}^{6} e_{\omega i}^2 \quad (10)$$

Hence, $\dot{V}(e_\omega)$ is calculated as:

$$\dot{V}(e_\omega) = \frac{1}{2}\sum_{i=1}^{6} e_{\omega i}\,\dot{e}_{\omega i} \quad (11)$$

$$\begin{aligned}
\dot{V}(e_\omega) &= e_{\omega 1}\,\dot{e}_{\omega 1} + e_{\omega 2}\,\dot{e}_{\omega 2} + e_{\omega 3}\,\dot{e}_{\omega 3} + e_{\omega 4}\,\dot{e}_{\omega 4} + e_{\omega 5}\,\dot{e}_{\omega 5} + e_{\omega 6}\,\dot{e}_{\omega 6} \\
&= e_{\omega 1}[ae_{\omega 3} - be_{\omega 1} - k_1 u_1] + e_{\omega 2}[ae_{\omega 4} - be_{\omega 2} - k_2 u_2] \\
&\quad + e_{\omega 3}[-ae_{\omega 3} + \omega_{1d}\omega_{5d} - k_3\omega_{1r}\omega_{5r} + 2ae_{\omega 3} - \omega_{2d}\omega_{6d} + k_3\omega_{2r}\omega_{6r} - k_3 u_3] + e_{\omega 4}[-ae_{\omega 4} + \omega_{1d}\omega_{6d} \\
&\quad - k_4\omega_{1r}\omega_{6r} + 2ae_{\omega 4} - \omega_{2d}\omega_{5d} - k_4\omega_{2r}\omega_{5r} - k_4 u_4] \\
&\quad + e_{\omega 5}[-de_{\omega 5} - c[\omega_{1d}\omega_{3d} - k_5\omega_{1r}\omega_{3r} - \omega_{2d}\omega_{4d} + k_5\omega_{2r}\omega_{4r}] - k_5 u_5] + e_{\omega 6}[-de_{\omega 6} - c[\omega_{1d}\omega_{4d} - \\
&\quad k_6\omega_{1r}\omega_{4r} - \omega_{2d}\omega_{3d} - k_6\omega_{2r}\omega_{3r}] - k_6 u_6]
\end{aligned}$$

Using the defined control inputs (9), $\dot{V}(e_\omega)$ is obtained as:

$$\dot{V}(e_\omega) = -be_{\omega 1}^2 - be_{\omega 2}^2 - ae_{\omega 3}^2 - ae_{\omega 4}^2 - de_{\omega 5}^2 - de_{\omega 6}^2 \quad (12)$$

Since $\dot{V}(e_\omega) < 0 \ \forall \ \omega_i \neq 0$, then using Lyapunov stability theorem, the errors $e_i \to 0$ as $t \to \infty$. Therefore, the states of drive (2) and response (3) hyperchaotic systems are synchronised. Further, $V(e_\omega)$ is radially unbounded, i.e. $V(e_\omega) \to \infty$ as $t \to \infty$, then it can be concluded that the error dynamics (7) is globally asymptotically stable.

3.2. MATLAB Simulation of Projective Synchronisation between Hyperchaotic Systems

The simulation is performed in MATLAB environment using ode45 solver. The projective synchronisation of state variables between the drive (5) and response (6) hyperchaotic systems is shown in Fig. 1. It is apparent that states of response system is synchronised with the drive system at the different projective factors. The initial conditions for drive system (5) is considered as $[\omega_{1d}(0) \ \omega_{2d}(0) \ \omega_{3d}(0) \ \omega_{4d}(0) \ \omega_{5d}(0) \ \omega_{6d}(0)]^T = [0.08 \ 0.02 \ 0.08 \ 0.08 \ 0.02 \ 0.08]^T$ and $[\omega_{1r}(0) \ \omega_{2r}(0) \ \omega_{3r}(0) \ \omega_{4r}(0) \ \omega_{5r}(0) \ \omega_{6r}(0)]^T = [1 \ 2 \ 1 \ 2 \ 1 \ 2]^T$ is for response system (6). Projective factors are considered as $k_1 = 2$, $k_2 = -1$, $k_3 = 2$, $k_4 = -1$, $k_5 = 1$ and $k_6 = -1$ where positive and negative values of k_i denotes the synchronisation and anti-synchronisation. In Fig. 2, the projective synchronisation errors are shown and ensures the convergence of error in less settling time (approx 0.3 seconds), i.e. synchronisation between drive and response chaotic systems is achieved successfully.

4. ENCRYPTION OF MESSAGE SIGNAL USING COMPLEX STATE VARIABLES-BASED HYPERCHAOTIC SYSTEM

In this section, first the traditional additive or masking technique is introduced and then two new encryption techniques for secure communication are illustrated and analysed.

The communication system has to be more reliable and secured to avoid data breaching. In the area of secure communication, chaos-based communication techniques have been receiving significant attention in the past few decades due to complex, random natures of chaotic systems. The well known encryption techniques are available in the literature as masking

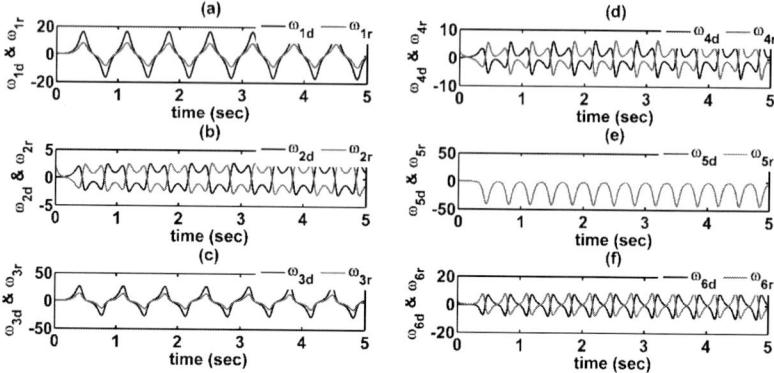

Figure 1. Time response of projective synchronised states of drive (5) and response (6) hyperchaotic systems.

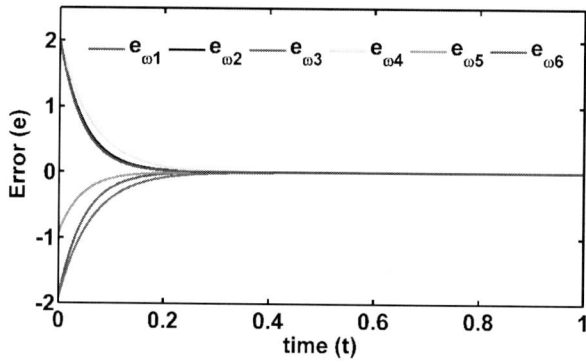

Figure 2. Projective synchronisation error between the drive (5) and response (6) hyperchaotic systems.

scheme, modulation scheme and chaotic switching (Please refer subsection 1.2). Encryption of a message signal is done to transmit a confidential data/information without any loss or breach of information.

4.1. Masking Technique for Secure Communication

The block diagram of secure communication using masking technique is given in Fig. 3.

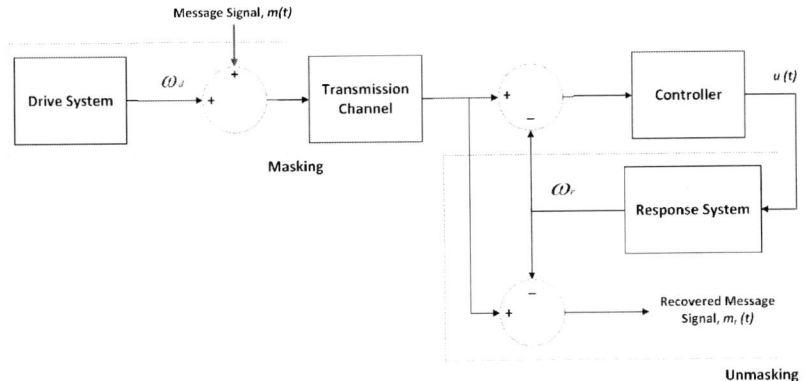

Figure 3. Block diagram of masking technique for secure communication.

Let message signal $m(t)$ be added in the first state of hyperchaotic drive system (5). The modified hyperchaotic drive system is written as:

$$\begin{cases} \dot{\omega}_{1d} = a\omega_{3d} - b(\omega_{1d} + m(t)) \\ \dot{\omega}_{2d} = a\omega_{4d} - b\omega_{2d} \\ \dot{\omega}_{3d} = \omega_{1d}\omega_{5d} - \omega_{2d}\omega_{6d} + a\omega_{3d} \\ \dot{\omega}_{4d} = \omega_{1d}\omega_{6d} + \omega_{2d}\omega_{5d} + a\omega_{4d} \\ \dot{\omega}_{5d} = -d\omega_{5d} - c\omega_{1d}\omega_{3d} + c\omega_{2d}\omega_{4d} \\ \dot{\omega}_{6d} = -d\omega_{6d} - c\omega_{1d}\omega_{4d} - c\omega_{2d}\omega_{3d} \end{cases} \quad (13)$$

The response system is considered as in (6). In order to achieve communication between the hyperchaotic drive and response systems, an appropriate controller is required at the receiver end. The control inputs to achieve the synchronisation between modified drive system (13) and response system (6) are designed as:

$$\begin{cases} u_1 = \dfrac{1}{k_1}[ae_{\omega 3} - bm(t)] \\ u_2 = \dfrac{1}{k_2}[ae_{\omega 4}] \\ u_3 = \dfrac{1}{k_3}[\omega_{1d}\omega_{5d} - k_3\omega_{1r}\omega_{5r} + 2ae_{\omega 3} - \omega_{2d}\omega_{6d} + k_3\omega_{2r}\omega_{6r}] \\ u_4 = \dfrac{1}{k_4}[\omega_{1d}\omega_{6d} - k_4\omega_{1r}\omega_{6r} + 2ae_{\omega 4} - \omega_{2d}\omega_{5d} + k_4\omega_{2r}\omega_{5r}] \\ u_5 = \dfrac{-c}{k_5}[\omega_{1d}\omega_{3d} - k_5\omega_{1r}\omega_{3r} - \omega_{2d}\omega_{4d} + k_5\omega_{2r}\omega_{4r}] \\ u_6 = \dfrac{-c}{k_6}[\omega_{1d}\omega_{4d} - k_6\omega_{1r}\omega_{4r} + \omega_{2d}\omega_{3d} - k_6\omega_{2r}\omega_{3r}] \end{cases} \quad (14)$$

The Lyapunov function candidate is considered as in (10). Using the above control inputs (14) and error dynamics (8), the rate of Lyapunov energy function $\dot{V}(e_\omega)$ is obtained as:

$$\dot{V}(e_\omega) = -be_{\omega 1}^2 - be_{\omega 2}^2 - ae_{\omega 3}^2 - ae_{\omega 4}^2 - de_{\omega 5}^2 - de_{\omega 6}^2 \quad (15)$$

Hence from (15), $\dot{V}(e_\omega) < 0 \ \forall \ \omega_i \neq 0$, therefore by using Lyapunov stability theorem, the modified drive and response hyperchaotic systems are synchronised and the transmission of message/information signal can be accomplished. The simulation results are shown in Figs. 6(a), 7, 10 and 13.

4.2. Diffie-Hellman Key Exchange - Masking Technique for Secure Communication

Diffie-Hellman (D-H) key exchange technique is a method of digital encryption that uses numbers raised to specific powers to generate decryption keys based on the components that are never transmitted directly, thereby mathematically overwhelming the function of a desired code breaker [46]. The D-H key exchange algorithm is summarized as follows:

- **Step 1:** Transmitter and Receiver have the pre-decided public keys (say 23 and 9).

- **Step 2:** Transmitter selects a private key (say 4) and Receiver selects a private key (say 3).

- **Step 3:** Transmitter and Receiver compute the public values as:
  Transmitter : $x = (9^4 \bmod 23) = (6561 \bmod 23) = 6$
  Receiver : $x = (9^3 \bmod 23) = (729 \bmod 23) = 16$

- **Step 4:** Transmitter and Receiver exchanges public keys as 6 and 16, respectively.

- **Step 5:** Transmitter and Receiver re-compute the keys as:
  Transmitter : $x = (9^{16} \bmod 23) = (65536 \bmod 23) = 9$
  Receiver : $x = (9^6 \bmod 23) = (216 \bmod 23) = 9$

- **Step 6:** 9 is the '*secret key*'.

The above algorithm is explored further with chaos based secure communication. The block diagram of D-H Key exchange based chaos communication technique is shown in Fig. 4. Let the secured key obtained via

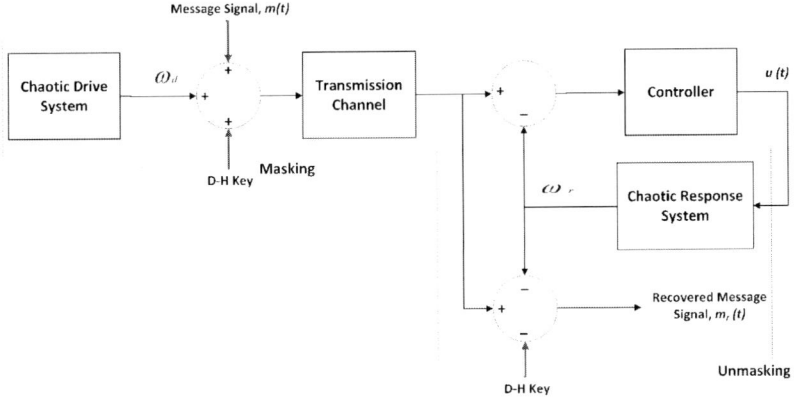

Figure 4. Block diagram of Diffie-Hellman key exchange-masking technique for secure communication.

D-H key exchange technique be added along with the sum of message signal $m(t)$ and the first state of hyperchaotic drive system (5). Based on the modified technique, the transmitted signal is defined as $(\omega_{1d} + m(t) + key)$

and the hyperchaotic drive system is obtained as:

$$\begin{cases} \dot{\omega}_{1d} = a\omega_{3d} - b(\omega_{1d} + m(t) + key) \\ \dot{\omega}_{2d} = a\omega_{4d} - b\omega_{2d} \\ \dot{\omega}_{3d} = \omega_{1d}\omega_{5d} - \omega_{2d}\omega_{6d} + a\omega_{3d} \\ \dot{\omega}_{4d} = \omega_{1d}\omega_{6d} + \omega_{2d}\omega_{5d} + a\omega_{4d} \\ \dot{\omega}_{5d} = -d\omega_{5d} - c\omega_{1d}\omega_{3d} + c\omega_{2d}\omega_{4d} \\ \dot{\omega}_{6d} = -d\omega_{6d} - c\omega_{1d}\omega_{4d} - c\omega_{2d}\omega_{3d} \end{cases} \quad (16)$$

The response system is considered as in (6). In order to achieve communication between the hyperchaotic drive and response systems, the control inputs for synchronisation between modified drive system (16) and response system (6) are further designed as:

$$\begin{cases} u_1 = \frac{1}{k_1}[ae_{\omega 3} - bm(t) - key \times b\omega_{1d} + bk_1\omega_{1r} + be_{\omega_1}] \\ u_2 = \frac{1}{k_2}[ae_{\omega 4}] \\ u_3 = \frac{1}{k_3}[\omega_{1d}\omega_{5d} - k_3\omega_{1r}\omega_{5r} + 2ae_{\omega 3} - \omega_{2d}\omega_{6d} + k_3\omega_{2r}\omega_{6r}] \\ u_4 = \frac{1}{k_4}[\omega_{1d}\omega_{6d} - k_4\omega_{1r}\omega_{6r} + 2ae_{\omega 4} - \omega_{2d}\omega_{5d} + k_4\omega_{2r}\omega_{5r}] \\ u_5 = \frac{-c}{k_5}[\omega_{1d}\omega_{3d} - k_5\omega_{1r}\omega_{3r} - \omega_{2d}\omega_{4d} + k_5\omega_{2r}\omega_{4r}] \\ u_6 = \frac{-c}{k_6}[\omega_{1d}\omega_{4d} - k_6\omega_{1r}\omega_{4r} + \omega_{2d}\omega_{3d} - k_6\omega_{2r}\omega_{3r}] \end{cases} \quad (17)$$

Using the above control inputs (17) and error dynamics (8), the rate of Lyapunov energy function $\dot{V}(e_\omega)$ is obtained as:

$$\dot{V}(e_\omega) = -be_{\omega 1}^2 - be_{\omega 2}^2 - ae_{\omega 3}^2 - ae_{\omega 4}^2 - de_{\omega 5}^2 - de_{\omega 6}^2 \quad (18)$$

From (18), $\dot{V}(e_\omega) < 0 \; \forall \; \omega_i \neq 0$, therefore using Lyapunov stability theorem, the modified drive and response systems are synchronised and the message/information signal is transmitted. The simulation results are shown in Figs. 6(b), 8, 11 and 14.

4.3. Modulation-Masking Technique for Secure Communication

In this approach, the modulation and masking techniques are cascaded and this hybrid technique is termed as mod-mask technique. The block diagram

Projective Synchronisation of Complex Valued Hyperchaotic ... 39

of mod-mask technique is shown in Fig. 5.

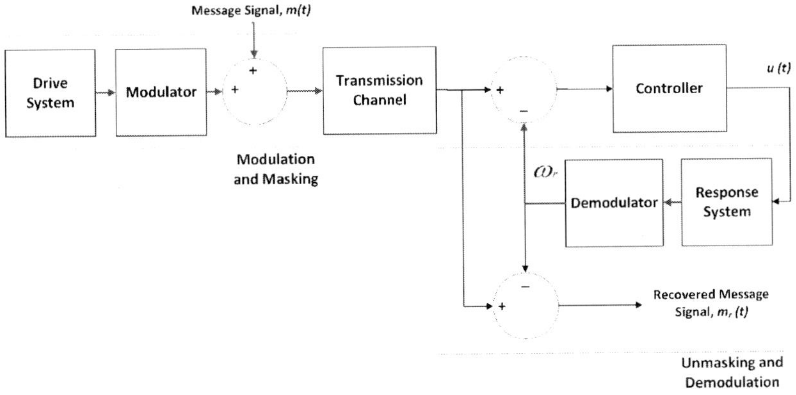

Figure 5. Block diagram of mod-mask technique for secure communication.

Let message signal, $m(t)$ be modulated first then the modulated signal is masked with the first state of hyperchaotic drive system (5). The transmitted signal in this technique is defined as $(\omega_{1d}m(t) + \omega_{1d})$ and the drive hyperchaotic system is modified as:

$$\begin{cases} \dot{\omega}_{1d} = a\omega_{3d} - b(\omega_{1d}m(t) + \omega_{1d}) \\ \dot{\omega}_{2d} = a\omega_{4d} - b\omega_{2d} \\ \dot{\omega}_{3d} = \omega_{1d}\omega_{5d} - \omega_{2d}\omega_{6d} + a\omega_{3d} \\ \dot{\omega}_{4d} = \omega_{1d}\omega_{6d} + \omega_{2d}\omega_{5d} + a\omega_{4d} \\ \dot{\omega}_{5d} = -d\omega_{5d} - c\omega_{1d}\omega_{3d} + c\omega_{2d}\omega_{4d} \\ \dot{\omega}_{6d} = -d\omega_{6d} - c\omega_{1d}\omega_{4d} - c\omega_{2d}\omega_{3d} \end{cases} \quad (19)$$

The response system for mod-mask technique is considered as in (6). In order to achieve communication, the control inputs for synchronisation between modified drive system (16) and response system (6) are further designed as:

$$\begin{cases} u_1 = \dfrac{1}{k_1}[ae_{\omega 3} - b\omega_{1d}(m(t)+1) + bk_1\omega_{1r} + be_{\omega 1}] \\ u_2 = \dfrac{1}{k_2}[ae_{\omega 4}] \\ u_3 = \dfrac{1}{k_3}[\omega_{1d}\omega_{5d} - k_3\omega_{1r}\omega_{5r} + 2ae_{\omega 3} - \omega_{2d}\omega_{6d} + k_3\omega_{2r}\omega_{6r}] \\ u_4 = \dfrac{1}{k_4}[\omega_{1d}\omega_{6d} - k_4\omega_{1r}\omega_{6r} + 2ae_{\omega 4} - \omega_{2d}\omega_{5d} + k_4\omega_{2r}\omega_{5r}] \\ u_5 = \dfrac{-c}{k_5}[\omega_{1d}\omega_{3d} - k_5\omega_{1r}\omega_{3r} - \omega_{2d}\omega_{4d} + k_5\omega_{2r}\omega_{4r}] \\ u_6 = \dfrac{-c}{k_6}[\omega_{1d}\omega_{4d} - k_6\omega_{1r}\omega_{4r} + \omega_{2d}\omega_{3d} - k_6\omega_{2r}\omega_{3r}] \end{cases} \quad (20)$$

Using the above control inputs (20) in the error dynamics (8), the derivative of Lyapunov function candidate $\dot{V}(e_\omega)$ is obtained as:

$$\dot{V}(e_\omega) = -be_{\omega 1}{}^2 - be_{\omega 2}{}^2 - ae_{\omega 3}{}^2 - ae_{\omega 4}{}^2 - de_{\omega 5}{}^2 - de_{\omega 6}{}^2 \quad (21)$$

From (21), $\dot{V}(e_\omega) < 0 \ \forall \ \omega_i \neq 0$, hence by using Lyapunov stability theorem, the modified drive (19) and response (6) systems are synchronised and the message/information signal is transmitted successfully. The simulation results for mod-mask technique are shown in Figs. 6(c), 9, 12 and 15.

5. ANALYSIS OF PROPOSED ENCRYPTION TECHNIQUE

The simulation is performed for the proposed D-H key exchange-masking and mod-mask hybrid encryption techniques as well as the traditional masking technique and comparison is achieved. Simulation is done in MATLAB environment using ode45 solver. The initial conditions and all the parameters values are kept same for all the three techniques. The synchronisation between the drive and response hyperchaotic systems is shown in Fig. 6 and confirms that the synchronisation between ω_{1d} state of hyperchaotic drive system and ω_{1r} state of hyperchaotic response system is achieved for all three techniques. In Figs. 7, 8 and 9, the convergence of average synchronisation error is taking approximately 0.25 seconds for all

Projective Synchronisation of Complex Valued Hyperchaotic ... 41

techniques. The average synchronisation error is defined as

$$e_{avg} = \frac{(e_{\omega_1} + e_{\omega_2} + e_{\omega_3} + e_{\omega_4} + e_{\omega_5} + e_{\omega_6})}{6}.$$

The message and recovered signals for each technique are shown in Figs. 10, 11 and 12 wherein it can be observed that the message signal is recovered successfully with very less convergence time of error between the message signal $m(t)$ and response signal $m_r(t)$ for the three techniques. The transmitted signal and its power spectrum for each techniques are shown in Figs. 13, 14 and 15 which shows the random or chaotic behaviour of transmitted signal (in left) and the power of the transmitted signals (in right) for each techniques. Fig. 14 shows that the amplitude of transmitted signal and its power spectrum under mod-mask technique is better among all the three techniques.

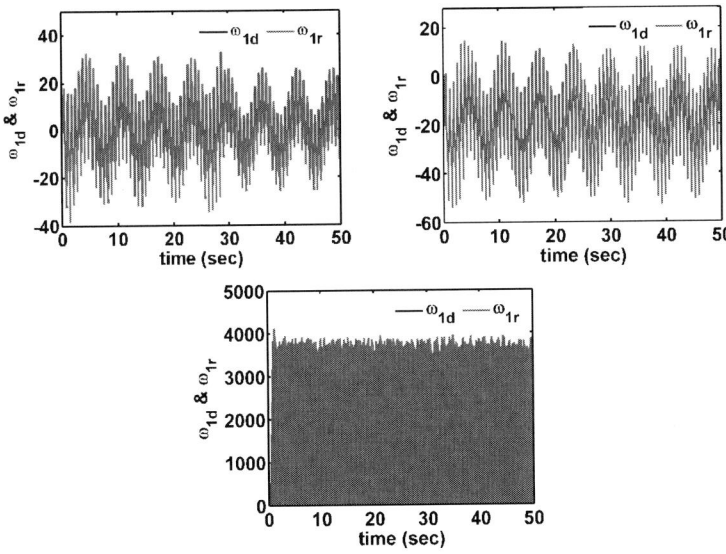

Figure 6. Time response of synchronisation of ω_{1d} state of drive and ω_{1r} state of response systems via (a) masking technique, (b) D-H key exchange-masking technique and (c) mod-mask technique.

It is apparent from Fig. 14 that the randomness of transmitted signals using D-H key exchange-masking and mod-mask techniques are better

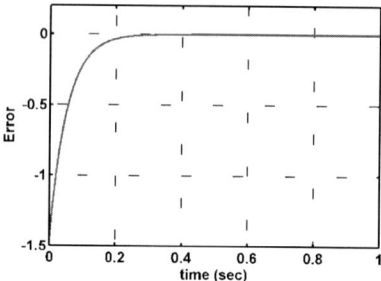

Figure 7. Average synchronisation error response for masking technique.

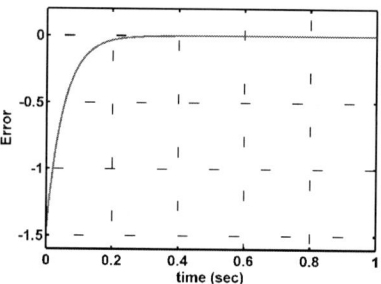

Figure 8. Average synchronisation error response for D-H key exchange-masking technique.

than the traditional masking technique. Moreover, mod-mask technique based random behaviour of transmitted signal is superior than the other two techniques. To justify the same, the power spectrum of the transmitted signal is also plotted and corresponds that the mod-mask encryption technique has better power strength over the large range of frequency.

Moreover, the analytical results for the D-H encryption techniques seems similar of the traditional masking technique however the algorithm for exchanging the secure key makes it more robust. Hence, both the proposed encryption techniques ensures more robust and secure communication.

Projective Synchronisation of Complex Valued Hyperchaotic ... 43

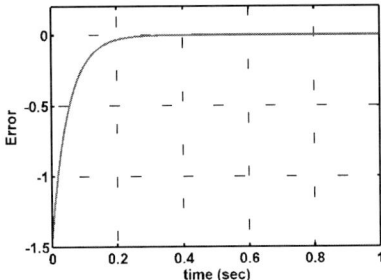

Figure 9. Average synchronisation error response for mod-mask technique.

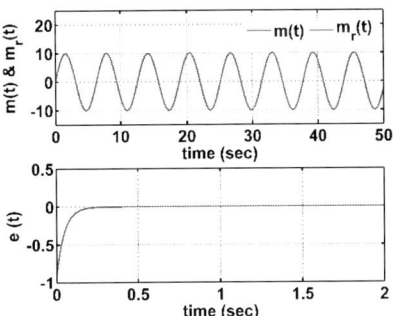

Figure 10. Time response and error response for message signal & recovered signal using masking technique.

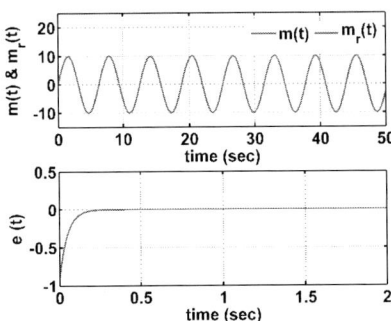

Figure 11. Time response and error response between message signal & recovered signal for D-H key exchange-masking technique.

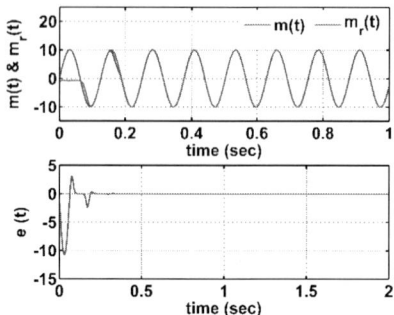

Figure 12. Time response of message signal & recovered signal and error response for mod-mask technique.

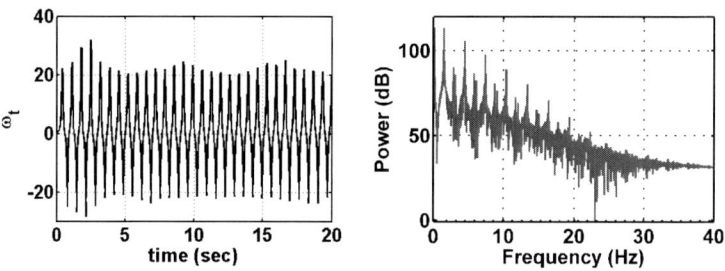

Figure 13. Transmitted signal and its power spectrum in the masking technique.

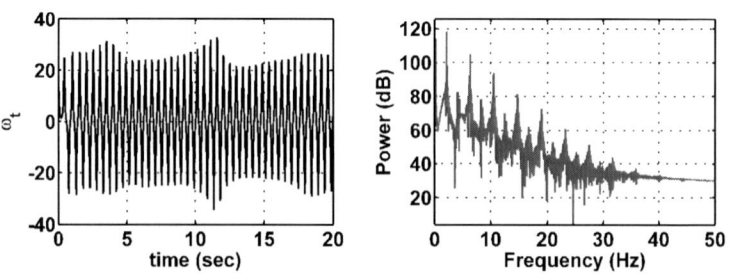

Figure 14. Shows transmitted signal and its power spectrum in D-H key exchange-masking technique.

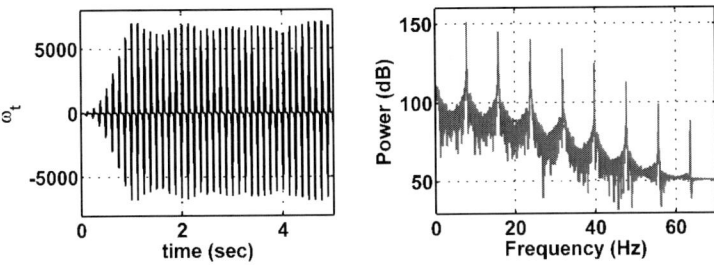

Figure 15. Transmitted signal and its power spectrum in the mod-mask technique.

CONCLUSION

The projective synchronisation between the derived and identical complex state variables based hyperchaotic systems is achieved using nonlinear active control technique. The synchronisation between hyperchaotic systems is achieved in very less settling time having different projective synchronisation factors. The global asymptotic stability under the synchronisation of the hyperchaotic system is ensured using appropriate design of control inputs. The required asymptotic stability is derived using Lyapunov stability theorem. The designed control technique for projective synchronisation is effective and easy to implement.

Further, two hybrid encryption (D-H key exchange-masking and mod-mask) techniques are proposed for secure communication using the complex state variables based hyperchaotic system. The proposed encryption techniques are compared with the traditional additive masking technique. The advantages of D-H key exchange-masking and mod-mask techniques are:

- The proposed encryption techniques are better than the traditional masking technique.

- Mod-mask technique is superior than the D-H key exchange-masking technique and traditional masking technique.

- D-H key exchange-masking technique is robust due to randomly generated key.

REFERENCES

[1] Pecora L. M. and Carrol T. L., "Synchronisation in chaotic system", *Physical Review Letters*, vol. 64, no. 8, pp. 821-825, 1990.

[2] Yurkon G. T., *"Introduction to chaos and it's real world applications"*, Cleveland State University, May 28, 1997.

[3] Saha P., Banerjee S. and Chowdhury A. R., "Chaos, signal communication and parameter estimation", *Physics Letters*, vol. 326, no. 2, pp. 133-139, 2004.

[4] Kumar R. R., Sampath A. and Indumathi P., "Improving security of communication systems using chaos", *Indian Journal of Science and Technology*, vol. 4, no. 5, pp. 561-565, May 2011.

[5] Chen G. and Dong X., *From chaos to order: methodologies, perspectives and applications*, World Scientific, Singapore, 1998, 311-387.

[6] Rauf K. and Yasir M., "Chaos based optical communication", *International Journal of Computer and Communication Engineering*, vol. 2, no. 2, pp. 97-100, March 2013.

[7] Rossler O. E., "An equation for hyperchaos," *Physics Letters A*, vol. 71, no. 2, pp. 155-157, 1979.

[8] Cenys A., Tamasevicius A. and Lindberg E., "Hyperchaos in coupled Colpitts oscillators," *Chaos, Solitons and Fractals*, vol. 17, no. 2, pp. 349-353, 2003.

[9] Li Y., Tang W. and Chen G., "Generating hyperchaos via state feedback control," *International Journal of Bifurcation and Chaos*, vol. 15, no. 10, pp. 3367-3375, 2005.

[10] Wang G., Zhang X., Zheng Y. and Li Y., "A new modified hyperchaotic Lu system," *Physica A: Statistical Mechanics and Its Applications*, vol. 371, no. 2, pp. 260-272, 2006.

[11] Barboza R., "Dynamics of a hyperchaotic Lorenz system," *International Journal of Bifurcation and Chaos*, vol. 17, no. 12, pp. 4285-4294, 2007.

[12] Thomas R., Basios V., and Rssler O. E., "Hyperchaos of arbitrary order generated by a single feedback circuit and the emergence of chaotic walks," *Chaos: An Interdisciplinary Journal of Nonlinear Science*, vol. 14, no.3, pp. 669-674, 2004.

[13] Wang J. and Chen Z., "A novel hyperchaotic system and its complex dynamics," *International Journal of Bifurcation and Chaos*, vol. 18, pp. 3309-3324, 2008.

[14] Vaidyanathan S., Pham V.T. and Volos C.K., "A 5-D hyperchaotic Rikitake dynamo system with hidden attractors," *European Physical Journal Special Topics*, vol. 224, no. 8, pp. 1575-1592, 2015.

[15] Liu Y. J., Yang Q. G. and Pang G. P., "A hyperchaotic system from the Rabinovich system," *Journal of Computer Applied Mathematics*, vol. 234, pp. 101-113, 2010.

[16] Mahmoud E. E. and Abood F. S., "A new nonlinear chaotic complex model and its complex antilag synchronisation," *Complexity*, vol. 2017, pp. 1-13, 2017.

[17] Yadav V. K., Kumar R., Leung A. Y. T. and Das S., "Dual phase and dual anti-phase synchronisation of fractional order chaotic systems in real and complex variables with uncertainties," *Chinese Journal of Physics*, vol. 57, pp. 282-308, 2019.

[18] Sun B., Li M. and Liu J., "The characteristics and self-time-delay synchronisation of two-time-delay complex Lorenz system;" *Journal of the Franklin Institute*, vol 356, pp. 334-350, 2019.

[19] Moghtadaei M. and Golpayegani M. R. H., "Complex dynamic behaviors of the complex Lorenz system," Scientia Iranica, vol. 19, pp. 733-738, 2012.

[20] Yadav V. K., Prasad G., Srivastava M. and Das S., "Combination-combination phase synchronisation among non-identical fractional order complex chaotic systems via nonlinear control," *International Journal of Dynamics and Control*, vol. 7, pp. 330-340, 2019.

[21] Mahmoud E. E. and Adwani M. A., "Dynamical behaviors, control and synchronisation of a new chaotic model with complex variables and cubic nonlinear terms," *Results in Physics*, vol. 7, pp. 1346-1356, 2017.

[22] Elkholy M. M., Hennawy H. M. E. and Elkouny A., "Simulation and digital implementation of eight dimensional hyper chaotic system for secured chaotic communication," *International Journal of Engineering Research & Technology*, vol. 04, no. 03, 482-486, 2015.

[23] Alsafasfeh Q. H. and Arni M. S., "A new chaotic behavior from Lorenz and Rossler systems and its electronic circuit implementation," *Circuits and Systems*, vol. 02, pp. 101-105, 2011.

[24] Yang T. and Chua L., "Secure Communication via chaotic parameter modulation", *IEEE Transactions on circuits and systems: Fundamental Theory and applications*, vol. 43, no. 9, pp. 817-819, September 1996.

[25] Angeli A. D., Genesio R. and Tesi A., "Dead-beat chaos synchronisation in discrete-time systems", *IEEE Transactions on circuits and systems: Fundamental Theory and applications*, vol. 42, no. 1,pp. 54-56, 1995.

[26] Chen S. and Lu J., "Synchronisation of an uncertain unified chaotic system via adaptive control", *Chaos, Solitons and Fractals*, vol. 14, no. 4, pp. 643-647, 2002.

[27] Xue C., Jiang N., Lv Y. and Qiu Kun, "Secure key distribution based on dynamic chaos synchronisation of cascaded semiconductor laser systems", *IEEE Transactions on Communications*, vol. 65, pp. 312-319, 2017.

[28] Li C., Sprott J. C., Akgul A. and Zhao Yibo, "A new chaotic oscillator with free control", *Chaos: An Interdisciplinary Journal of Nonlinear Science*, vol. 27, pp. 083101-083106, 2017.

[29] Wen G., Chen M. Z. Q. and Yu X., "Event-triggered master-slave synchronisation with sampled-data communication", *IEEE Transactions on Circuits and Systems*, vol. 63, no. 3, pp. 304-308, 2016.

[30] Souaia M. A., Trabelsi H. and Saad K. B., "Synchronisation of the Liu chaotic system and its application in secure communication", *2017 International Conference on Control, Automation and Diagnosis (ICCAD), Hammamet*, vol. 17, pp. 434-438, 2017.

[31] Aromataris G. and Lodi V. A., "Assessing Security of Chaos Communications Against Eavesdropping by Pump Detection", *IEEE Journal of Quantum Mechanics*, vol. 55, no. 2, pp. 101-104, 2019.

[32] Singh P. P., Singh J. P., Borah M. and Roy B. K., "On the construction of a new chaotic system", *IFAC-PapersOnLine*, vol. 49, no.1, pp. 522-525, 2016.

[33] Wang B. and Zhong S. M., "Observer-based control on a chaotic system with unknowns and uncertainties", *International Journal for Light and Electron Optics*, vol. 137, pp. 167-174, 2017.

[34] Li Z., Li K., Wen C. and Soh Y. C., "A new chaotic secure communication system", *IEEE Transactions on Communication*, vol. 51, no. 8, pp. 1306-1312, 2003.

[35] Singh P. P. and Roy B. K., "Memristor based novel complex chaotic system and its projective synchronisation using nonlinear active control technique", *European Physical Journal Special Topics*, vol. 228, pp. 2197-2214, 2019.

[36] Zaher A. A. and Rezq A. A., "On the design of chaos-based secure communication systems," *Communications in Nonlinear Science and Numerical Simulation*, vol. 16, no. 9, pp. 3721-3737, 2011.

[37] Tam W. M., Lau F. C. M., Tse C. K., *Digital communication with chaos multiple access techniques and performance*, Elsevier, 2007.

[38] Quyen N. X., Van Yem V. and Hoang T. M., "Chaotic modulation based on the combination of CPPM and CPWM" *Proceedings of the Joint INDS'11 & ISTET'11, Klagenfurt*, pp. 1-6, 2011.

[39] Pandey A. and Dahate S., "Secure communication using a modified chaotic masking scheme employing Sprott94 case a chaotic attractor," *2014 2nd International Conference on Emerging Technology Trends in Electronics, Communication and Networking*, Surat, pp. 1-6, 2014.

[40] Alvarez G., Montoya F., Pastor G., and Romera M., "Breaking a secure communication scheme based on the phase synchronisation of chaotic systems," *Chaos: An Interdisciplinary Journal of Nonlinear Science*, vol. 14, no. 2, pp. 274, 2004.

[41] Li S., Alvarez G., Chen G. and Mou X., "Breaking a chaos-noise-based secure communication scheme," *Chaos: An Interdisciplinary Journal of Nonlinear Science*, vol. 15, no. 1, pp. 013703, 2005.

[42] Jinfeng H. and JingBo G., "Breaking a chaotic secure communication scheme," *Chaos: An Interdisciplinary Journal of Nonlinear Science*, vol. 18, no. 1, pp. 013121, 2008.

[43] Alvarez G., Montoya F., Pastor G., and Romera M., "Breaking parameter modulated chaotic secure communication system," *Chaos, Solitons & Fractals*, vol. 21, no. 4, pp. 783-787, 2004.

[44] Alvarez G., Montoya F., Pastor G., and Romera M., "Breaking projective chaos synchronisation secure communication using filtering and generalized synchronisation," *Chaos, Solitons & Fractals*, vol. 24, no. 3, pp. 775-783, 2005.

[45] Alvarez G. and Li S., "Breaking network security based on synchronized chaos," *Computer Communication*, vol. 27, no. 16, pp. 1679-1681, 2004.

[46] Diffie W. and Hellman M. E., "New Directions in Cryptography," *IEEE Transaction on Information Theory*, vol. 22, no. 6, pp. 105-111, 1976.

In: Nonlinear Systems
Editor: P. P. Singh

ISBN: 978-1-68507-660-3
© 2022 Nova Science Publishers, Inc.

Chapter 3

CHAOS CONTROL IN ELECTRIC VEHICLE USING CASCADED PI AND PID SLIDING MODE CONTROL TECHNIQUES

Ranjan Kumar[*], *Nathaneal Kharlukhi, Md. Tanweer Alam and Piyush Pratap Singh*
Department of Electrical Engineering,
National Institute of Technology Meghalaya, Shillong, India

ABSTRACT

This chapter develops and compares the nonlinear control of permanent magnet synchronous motor system using conventional Sliding Mode Control (SMC), PI SMC and PID SMC. The SMC method is well known for its invariant properties to certain internal parameter variations and external disturbances, guaranteeing perfect tracking performance despite parameters or model uncertainties. The robustness of conventional SMC can only be certified by selecting appropriate control gains. In contrast, the large gains will lead to the well-known chattering phenomenon, which can excite high-frequency dynamics. Thus, PI and PID SMC techniques are proposed to overcome the chattering phenomenon. Simulation results show the validity of the proposed control approach. Similarly, a cascaded PI SMC and PID

---
[*] Corresponding Author's E-mail: harikantranjan123@gmail.com.

SMC for speed control in electric vehicle (EV) is proposed and simulated in MATLAB to show the validity of the proposed control approaches. Numerical simulation results validate the theoretical approach and reveal the successful achievement of the objectives.

1. INTRODUCTION

We know the benefits of feedback control when the permanent magnet synchronous motor (PMSM) system has some model uncertainty or disturbance. The addition of control in the closed-loop dramatically improves the performance compared to open-loop control. It completely neglects disturbance and model uncertainty. In the control of the PMSM system, the proportional-integral (PI) control technique is still popular due to its simple design and implementation [1]. However, a practical PMSM system has large disturbances and uncertainties, which may come internally or externally, e.g., unmodelled dynamics, parameter variation, friction force, and load disturbances. It will be very difficult to limit these disturbances rapidly if adopting linear control methods like the PI control algorithm [2]. Therefore, many nonlinear control methods have been adopted to improve the control performances in systems with different disturbances and uncertainties, e.g., robust control, sliding-mode control (SMC), adaptive control, backstepping control, predictive control, intelligent control, etc. The SMC method is well known for its invariant properties to certain internal parameter variations and external disturbances in these nonlinear control methods. It can guarantee perfect tracking performance despite parameters or model uncertainties and has been successfully applied in many fields [3, 4].

The paper [5] applied the fuzzy sliding-mode approach to a six-phase induction machine. A hybrid terminal sliding-mode observer was proposed based on the non-singular terminal sliding mode and the high-order sliding mode for the rotor position and speed estimation in one PMSM control system [6]. In the paper [7], the performance of a sliding mode controller was studied using a hybrid controller applied to induction motors via sampled closed representations. The results were very conclusive regarding the effectiveness of the sliding-mode approach. A neuron-fuzzy sliding-mode controller applied to an induction machine can also be found in [8]. However, the robustness of SMC can only be guaranteed by the selection of large control gains, while the large gains will lead to the well-known chattering phenomenon, which can excite high-frequency dynamics. Thus, some approaches have been proposed to overcome chattering, such as continuation control, high-order sliding-mode method [9], and complementary sliding-mode method [10].

Electric vehicles (EVs) are advancing with concerns about protecting the environment and preserving energy sources; electric vehicles (EVs) are making advances. EV has many advantages, including smooth and quiet operation and high efficiency [11, 12]. Moreover, it plays a vital role in the advancement of fuel economy to satisfy the tightened emission standards. Excavation on EVs' power motive force system has been paid attention to in the automobile industries and academic research studies [12]. The power supply, drive train, motor, charger, and controller are the most significant parts of an EV system. A controller plays a decisive role in obtaining a high-performance EV with an optimal balance of traveling range per charge, maximum speed, and accelerated performance [13, 14]. EV systems are time-variant, e.g., operation parameters of the electric vehicle and the road condition are continuously varying. As a result, the control of EVs should be designed robustly and adaptively to improve the system on both dynamic and steady-state performances [14, 15]. At present, the short running distance per battery charge makes EVs' use limited [11]. The fact that EVs are "energy-management" machines makes their control special. Therefore, alongside controlling the performance of the EV, energy management of the batteries on the EV is of great importance.

Motivated by the above literature, the following objectives are considered in this study.

1. Chaos control in the PMSM using PI and PID sliding mode control techniques and comparison.
2. Electric vehicle speed control using cascaded PI and PID sliding mode control techniques.

2. MODELING OF PERMANENT MAGNET SYNCHRONOUS MOTOR (PMSM)

The dynamics of a PMSM based on the d-q axis is modeled [15] and written as

$$\begin{cases} \frac{di_d}{dt} = (u_d - R_1 i_d + \omega L_q i_q)/L_d \\ \frac{di_q}{dt} = (u_q - R_1 i_q - \omega L_d i_d - \omega \varphi_r)/L_q \\ \frac{dy}{dx} = [n_p \varphi_r i_q + n_p(L_d - L_q) i_d i_q - T_L - \beta \omega]/J \end{cases} \quad (1)$$

By applying an affine transform $x = \lambda \tilde{x}$ and a time scaling transform $t = \tau \tilde{t}$, where $x = [i_d \; i_q \; \omega]^T$, $\tilde{x} = [\tilde{i}_d \; \tilde{i}_q \; \tilde{\omega}]^T$ represent the state vectors and

$$\lambda = \begin{bmatrix} \lambda_d & 0 & 0 \\ 0 & \lambda_q & 0 \\ 0 & 0 & \lambda_\omega \end{bmatrix} = \begin{bmatrix} \frac{L_q}{L_d} & 0 & 0 \\ 0 & \frac{\beta}{n_p \tau \varphi_r} & 0 \\ 0 & 0 & \frac{R_1}{L_q} \end{bmatrix},$$

The dynamics of the PMSM system in the dimensionless form is obtained as

$$\begin{cases} \frac{d\tilde{i}_d}{dt} = -\tilde{i}_d + \tilde{\omega}\tilde{i}_q + \tilde{u}_d \\ \frac{d\tilde{i}_q}{dt} = -\tilde{i}_q - \tilde{\omega}\tilde{i}_q + \gamma\tilde{\omega} + \tilde{u}_q \\ \frac{d\tilde{i}_\omega}{dt} = \sigma(\tilde{i}_q - \tilde{\omega}) + \epsilon \tilde{i}_d \tilde{i}_q - \tilde{T}_L \end{cases} \quad (2)$$

Using parameters $\gamma = -\frac{\varphi_r}{kL_q}$, $\sigma = \frac{\beta \tau}{J}$, $\tilde{u}_q = \frac{1}{R_1 k} u_q$, $\tilde{u}_d = \frac{1}{R_1 k} u_d$, $\varepsilon = \frac{n_p b_T{}^2 k^2 (L_d - L_q)}{J}$, $\tilde{T}_L = \frac{\tau^2}{J} T_L$, $b = \frac{L_q}{L_d}$ and $k = \frac{\beta}{n_p \tau \varphi_r}$ and $\frac{1}{\tau} = \frac{R_1}{L_q}$, the dynamics are further modified as

$$\begin{cases} \frac{\widetilde{di_d}}{d\tilde{t}} = -\widetilde{i_d} + \widetilde{\omega}\widetilde{i_q} + \widetilde{u_d} \\ \frac{\widetilde{di_q}}{d\tilde{t}} = -\widetilde{i_q} - \widetilde{\omega}\widetilde{i_q} + \gamma\widetilde{\omega} + \widetilde{u_q} \\ \frac{\widetilde{di_\omega}}{d\tilde{t}} = \sigma(\widetilde{i_q} - \widetilde{\omega}) - \widetilde{T_L} \end{cases} \quad (3)$$

The dynamics of PMSM show chaotic behaviour in [15] and maybe referred for detailed information.

3. Design and Stability of P, PI, and PID Sliding Mode Controllers for PMSM

The PI controller is the most popular variation, even more than full PID controllers. The value of the controller output $u(t)$ is fed into the system as the manipulated variable as:

$$u(t) = u_{bias} + K_p e(t) + K_i \int e(t)dt \quad (4)$$

The u_{bias} term is a constant that is typically set to the value of $u(t)$ when the controller is first switched from manual to automatic mode, this gives a "bump less" transfer if the error is zero when the controller is turned on. The two tuning values for a PI controller are the controller gains K_p and K_i. The value of K_p are a multiplier on the proportional error, and integral term and a higher value make the controller more aggressive at responding to errors away from the set point. The setpoint (SP) is the target value, and the process variable (PV) is the measured value that may deviate from the desired value. The error from the setpoint is the difference between the SP and PV and is defined as $e(t) = SP - PV$

The working principle behind a PID controller is that the proportional, integral, and derivative constants must be individually adjusted or tuned based on the difference between these values; a correction factor is calculated. Here are the three steps:

1. Proportional tuning involves correcting a target proportional to the difference. Thus, the target value is never achieved because as the difference approaches zero, so does the applied correction.
2. Integral tuning attempts to remedy this by effectively cumulating the error result from the "P" action to increase the correction factor. For example, if the oven remained below temperature, "I" would act to increase the head delivered. However, rather than stop heating when the target is reached. "I" attempts to drive the cumulative error to zero, resulting in an overshoot.
3. Derivative tuning minimizes this overshoot by slowing the correction factor applied as the target approaches.

3.1. Proportional Sliding Mode Control Design for PMSM

In this section, we will discuss the conventional/proportional SMC used as a tracking control problem for the speed control of PMSM. The control objective is to track the desired or reference signal. So, the error signal (e) between the actual and desired signals is defined as $e = x - d$, where d - Vector of desired signals and x − Vector of actual signals.

Let the tracking error is defined as

$$\begin{cases} e_1 = x_1 - d_1 \\ e_2 = x_2 - d_2 \\ e_3 = x_3 - d_3 \end{cases} \quad (5)$$

In the regulation control system, the Lyapunov candidate function is selected to be a function of the state dynamics and must be a positive definite. In this case, the objective that the tracking error $e(t)$ of the state is represented as the sliding surface s. Let the sliding surface be defined as

$$\begin{cases} s_1 = e_1 \\ s_2 = e_2 \\ s_3 = e_3 \end{cases} \quad (6)$$

The stability of SMC can be achieved based on the Lyapunov stability concept. According to Lyapunov stability concept, if a scalar function $V(x)$ exists with its continuous first-order derivative such that

(i) $V(x)$ Is positive definite
(ii) $\dot{V}(x)$ Is negative definite
(iii) $V(x) \to \infty$ as $|x| \to \infty$

Then the equilibrium at the origin is global asymptotically stable. Let the Lyapunov function candidate be defined as

$$V = \frac{1}{2}s^2 \qquad (7)$$

Which must be positive definite. Its first-order derivative can be expressed as $\dot{V} = s\dot{s} < 0 \; \forall \, s$, which needs to be negative definite. Now the first-order derivative of the above Equation then

$$\dot{s} = \dot{x} - \dot{d} + u_{eq}$$

where u_{eq} is the reaching control law. We need now to force $\dot{s}$ to go to zero as $0 = \dot{x} - \dot{d} + u_{eq}$ and equivalent control is obtained as $u_{eq} = \dot{d} - \dot{x}$ and written as

$$\begin{cases} u_{1eq} = \dot{d}_1 - \dot{x}_1 \\ u_{2eq} = \dot{d}_2 - \dot{x}_2 \\ u_{3eq} = \dot{d}_3 - \dot{x}_3 \end{cases} \qquad (8)$$

where $\dot{x}_1 = \frac{d\widetilde{i_d}}{dt}$, $\dot{x}_2 = \frac{d\widetilde{i_q}}{dt}$ and $\dot{x}_3 = \frac{d\widetilde{i_\omega}}{dt}$. Using the system dynamics (3) in the above Equation (8), the equivalent control law is written as

$$\begin{cases} u_{1eq} = \dot{d}_1 - \left(-\tilde{i}_d + \tilde{\omega}\tilde{i}_q + \tilde{u}_d\right) \\ u_{2eq} = \dot{d}_2 - \left(-\tilde{i}_q - \tilde{\omega}\tilde{i}_q + \gamma\tilde{\omega} + \tilde{u}_q\right) \\ u_{3eq} = \dot{d}_3 - \left(\sigma(\tilde{i}_q - \tilde{\omega}) - \tilde{T}_L\right) \end{cases} \quad (9)$$

The above Equation (2.8) ensures that the system trajectory will head to sliding surface s, $\dot{s} = 0$, $s \to$ constant, and that results $\dot{V} = s\dot{s} = 0$. But this is not sufficient for the stability condition because we need the trajectory to stay on the surface $s = 0$, which is the error signal $e = 0$. To achieve $\dot{V} < 0$, another term is necessary for control to ensure that the trajectory remains on the sliding surface for every s. This term is known as switching function as 'sign' function which is given by

$$sign(s) = \begin{cases} 1 & s > 0 \\ 0 & s = 0 \\ -1 & s < 0 \end{cases}$$

and

$$u_r = -\S\, sign(s) \quad (10)$$

where u_r −Switching control and $\S$ − Switching gain. Therefore, net control is defined as $u_{net} = u_{eq} + u_r$ and written as:

$$u_{net} = u_{eq} - \S\, sign(s) \quad (11)$$

Using Equation (11), the net control law for PMSM is written as:

$$\begin{cases} u_{1net} = \dot{d}_1 - \tilde{i}_d + \tilde{\omega}\tilde{i}_q + \tilde{u}_d - \S\, sign(s_1) \\ u_{2net} = \dot{d}_2 - \tilde{i}_q - \tilde{\omega}\tilde{i}_q + \gamma\tilde{\omega} + \tilde{u}_q - \S\, sign(s_2) \\ u_{3net} = \dot{d}_3 - \sigma(\tilde{i}_q - \tilde{\omega}) - \tilde{T}_L - \S\, sign(s_3) \end{cases} \quad (12)$$

Desired values of PMSM dynamics are considered as $d_1 = 17.21$, $d_2 = 1$, $d_3 = 0$ and simulated in MATLAB environment.

Chaos Control in Electric Vehicle using Cascaded PI and PID ... 57

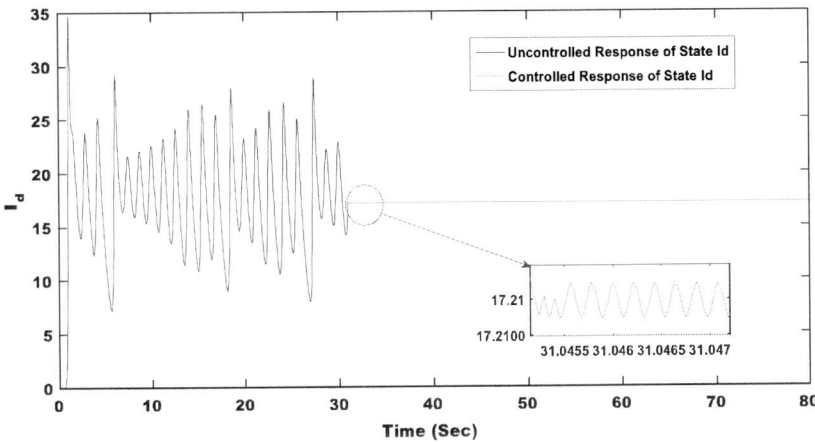

Figure 1. Controlled response of direct axis current (i_d).

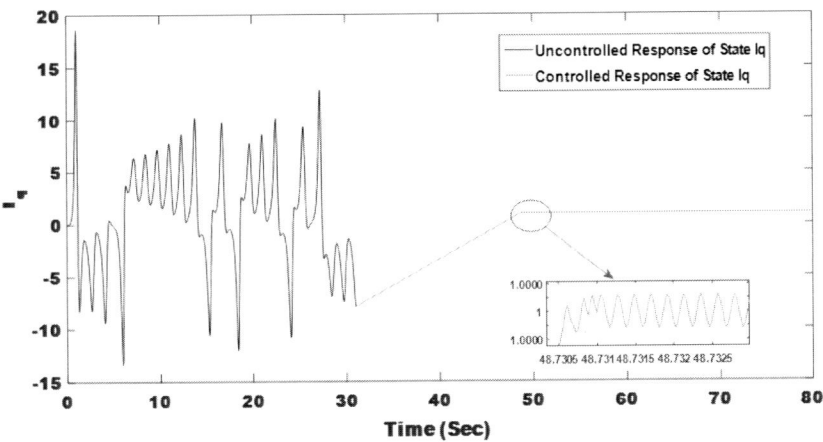

Figure 2. Controlled response of quadrature axis current (i_d) using SMC.

Figure 1 shows the chaotic response till 31 seconds and at 31 seconds, when SMC controller is applied, the chaotic behaviour of direct axis current (i_d) is controlled (refer 31 seconds to 80 seconds). In Figure 2, the chaotic behaviour is available till 31 seconds after 31 seconds, the chaotic behaviour of quadrature axis current (i_d) is controlled to its reference point once the SMC controller is applied. Similarly, the controlled behaviour of speed of PMSM is shown in Figure 3. The proportional SMC can control the chaotic behaviour, but a chattering effect in the steady-state needs to be eliminated using an appropriate control design.

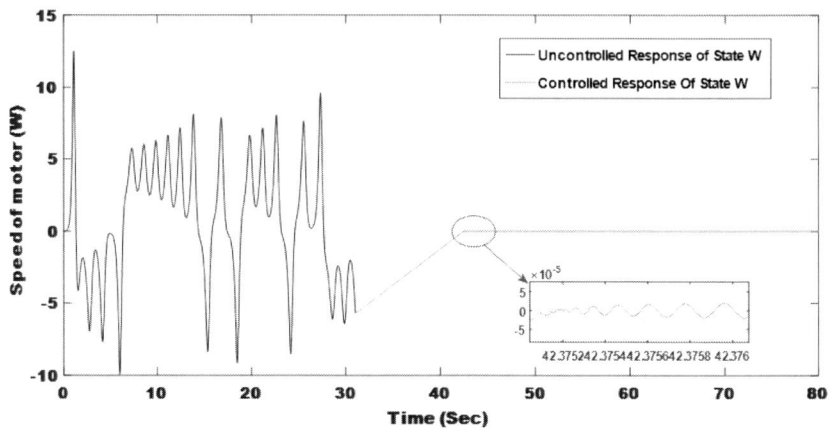

Figure 3. Controlled response of speed of the motor (ω).

3.2. Design and Stability of PI Sliding Mode Controller for PMSM

The PMSM dynamics can be represented in the state space form if $\tilde{\iota}_d = x_1, \tilde{\iota}_q = x_2$ and $\tilde{\omega} = x_3$ and written as:

$$\begin{cases} \dot{x}_1 = -x_1 + x_3 x_2 + U_d \\ \dot{x}_2 = -x_1 - x_3 x_2 + \gamma x_3 + U_q \\ \dot{x}_3 = \sigma(x_2 - x_3) - T_L \end{cases} \quad (13)$$

Let the tracking error be defined as:

$$\begin{cases} e_1 = x_1 - d_1 \\ e_2 = x_2 - d_2 \\ e_3 = x_3 - d_3 \end{cases} \quad (14)$$

Then the sliding surface is

$$\begin{cases} s_1 = e_1 + \int_0^t k_1\, e_1(\tau)d\tau \\ s_2 = e_2 + \int_0^t k_2\, e_2(\tau)d\tau \\ s_3 = e_3 + \int_0^t k_3\, e_3(\tau)d\tau \end{cases} \quad (15)$$

where the value of integral gains k_1, k_2 and k_3 are chosen 3, 5, and 8 are respectively. The sliding dynamics is written as:

$$\begin{cases} \dot{s}_1 = \dot{e}_1 + 3e_1 \\ \dot{s}_2 = \dot{e}_2 + 5e_2 \\ \dot{s}_3 = \dot{e}_3 + 8e_3 \end{cases} \quad (16)$$

The system operates in the sliding mode when $s_i = 0$ for $i = 1, 2, 3$. Therefore, the equivalent error dynamics is written as:

$$\begin{cases} \dot{e}_1 = \dot{x}_1 - \dot{d}_1 \\ \dot{e}_2 = \dot{x}_2 - \dot{d}_2 \\ \dot{e}_3 = \dot{x}_3 - \dot{d}_3 \end{cases} \quad (17)$$

The net control law is defined as $u_{net_i} = u_{eq_i} + u_{r_i}$ to ensure the occurrence of sliding motion, and becomes

$$\begin{cases} u_{net_1} = u_{eq_1} + u_{r_1} \\ u_{net_2} = u_{eq_2} + u_{r_2} \\ u_{net_3} = u_{eq_3} + u_{r_3} \end{cases} \quad (18)$$

The reaching control laws are defined as:

$$\begin{cases} u_{r_1} = -\zeta_1 sign(s_1) \\ u_{r_2} = -\zeta_2 sign(s_2) \\ u_{r_3} = -\zeta_3 sign(s_3) \end{cases} \quad (19)$$

The equivalent control laws are defined as:

$$\begin{cases} u_{eq_1} = -(x_1 - x_3x_2 - U_d + \dot{d}_1 - 3x_1 + 3d_1) \\ u_{eq_2} = -(x_1 + x_3x_2 - \gamma x_3 - U_q + \dot{d}_2 - 5x_2 + 5d_2) \\ u_{eq_3} = -(-\sigma(x_2 - x_3) + T_L + \dot{d}_3 - 8x_3 + 8d_3) \end{cases} \quad (20)$$

Using Equations (19) and (20), the net control laws in Equation (18) becomes

$$\begin{cases} u_{net_1} = x_1 - x_3x_2 - U_d + \dot{d}_1 - 3x_1 + 3d_1 - \zeta_1 sign(s_1) \\ u_{net_2} = x_1 + x_3x_2 - \gamma x_3 - U_q + \dot{d}_2 - 5x_2 + 5d_2 - \zeta_2 sign(s_2) \\ u_{net_3} = -\sigma(x_2 - x_3) + T_L + \dot{d}_3 - 8x_3 + 5d_3 - \zeta_3 sign(s_3) \end{cases}$$
(21)

To check the stability, Let the Lyapunov function candidate be defined as:

$$V(s) = \frac{1}{2}s_i^2 \text{ for } i = 1, 2, 3.$$

Assuming $V(s)$ continuously differentiable function to ensure a stable sliding surface. Then, the s_1 sliding surface has

$$\dot{V}(s_1) = s_1 \dot{s}_1$$
$$\dot{V}(s_1) = s_1(\dot{e}_1 + 3e_1)$$
$$\dot{V}(s_1) = s_1(-x_1 + x_3x_2 + U_d - \dot{d}_1 + 3x_1 - 3d_1 + u_{net_1})$$

Substituting the value of u_{net_1} from equation (21), we get

$$\dot{V}(s_1) = s_1(-\zeta_1 sign(s_1))$$
$$\dot{V}(s_1) = -s_1\zeta_1|sign(s_1)|$$

As per Lyapunov stability theory, $\dot{V}(s_1) < 0$. where $\dot{V}$ is negative definite iff ζ_1 is positive. Therefore, the error $e_1(t) \to 0$ at $t \to \infty$. As a result, the stability condition is satisfied $\forall s$.

Similarly, for surface s_2

$$\dot{V}(s_2) = s_2 \dot{s}_2$$
$$\dot{V}(s_2) = s_2(\dot{e}_2 + 5e_2)$$
$$\dot{V}(s_2) = s_2(-x_1 - x_3 x_2 + \gamma x_3 + U_q - \dot{d}_2 + 5x_2 - 5d_2 + u_{net_2})$$

By substituting the value of u_{net_2} from (21), $\dot{V}(s_2)$ is written as:

$$\dot{V}(s_2) = s_2(-\zeta_2 sign(s_2))$$
$$\dot{V}(s_2) = -s_2 \zeta_2 |sign(s_2)|$$

Since ζ_2 is positive, so $\dot{V}(s_2) < 0$ negative definite, which proves that the error $e_2(t) \to 0$ at $t \to \infty$. As a result, the stability condition is satisfied $\forall\ s$.

Similarly, for surface s_3

$$\dot{V}(s_3) = s_3 \dot{s}_3$$
$$\dot{V}(s_3) = s_3(\dot{e}_3 + 8e_3)$$
$$\dot{V}(s_3) = s_3(-\sigma(x_2 - x_3) + T_L + \dot{d}_3 - 8x_3 + 5d_3 + u_{net_3})$$

By substituting the value of u_{net_3} from equation (21), we get $\dot{V}(s_3)$ as

$$\dot{V}(s_3) = s_3(-\zeta_3 sign(s_3))$$
$$\dot{V}(s_3) = -s_3 \zeta_3 |sign(s_3)|$$

$\dot{V}(s_3) < 0$ since ζ_3 is positive, and it proves that the error $e_3(t) \to 0$ at $t \to \infty$. As a result, the stability condition is satisfied $\forall\ s$.

3.3. Design and Stability of PID Sliding Mode Controller for PMSM

Here, PID sliding mode control design is discussed for the defined PMSM dynamics (13). Assuming the same tracking error as described in (14), then the PID sliding surfaces are defined as:

$$\begin{cases} s_1 = e_1 + \int_0^t k_1 e_1(\tau) d\tau + D_1 \dfrac{d}{d_f} e_1 \\ s_2 = e_2 + \int_0^t k_2 e_2(\tau) d\tau + D_2 \dfrac{d}{d_f} e_2 \\ s_3 = e_3 + \int_0^t k_3 e_3(\tau) d\tau + D_3 \dfrac{d}{d_f} e_3 \end{cases} \quad (22)$$

where the values of integral gains k_1, k_2 and k_3 are kept the same as in the case of PI SMC, and the values of derivative gains $D_1, D_2,$ and D_3 are considered 15, 18, and 20, respectively, then the sliding dynamics is written as:

$$\begin{cases} \dot{s}_1 = \dot{e}_1 + 3e_1 + 15 \dfrac{d^2}{dt} e_1 \\ \dot{s}_2 = \dot{e}_2 + 5e_2 + 18 \dfrac{d^2}{dt} e_2 \\ \dot{s}_3 = \dot{e}_3 + 8e_3 + 20 \dfrac{d^2}{dt} e_3 \end{cases} \quad (23)$$

The system operates in the sliding mode when $s_i = 0$ for $i = 1, 2, 3$. Therefore, the error dynamics is written as:

$$\begin{cases} \dot{e}_1 = \dot{x}_1 - \dot{d}_1 \\ \dot{e}_2 = \dot{x}_2 - \dot{d}_2 \\ \dot{e}_3 = \dot{x}_3 - \dot{d}_3 \end{cases} \quad (24)$$

Chaos Control in Electric Vehicle using Cascaded PI and PID ...

The net control laws are defined in (25) to ensure the occurrence of sliding motion.

$$\begin{cases} u_{net_1} = u_{eq_1} - \zeta_1 sign(s_1) \\ u_{net_2} = u_{eq_2} - \zeta_2 sign(s_2) \\ u_{net_3} = u_{eq_3} - \zeta_3 sign(s_3) \end{cases} \quad (25)$$

where the equivalent control laws are obtained as:

$$\begin{cases} u_{eq_1} = x_1 - x_3 x_2 - U_d + \dot{d}_1 - 3x_1 + 3d_1 - 15\frac{d}{dt}(-x_1 + x_3 x_2 + U_d) \\ u_{eq_2} = x_1 + x_3 x_2 - \gamma x_3 - U_q + \dot{d}_2 - 5x_2 + 5d_2 - 18\frac{d}{dt}(-x_1 - x_3 x_2 + \gamma x_3 + U_q) \\ u_{eq_3} = -\sigma(x_2 - x_3) + T_L + \dot{d}_3 - 8x_3 + 5d_3 - 20\frac{d}{dt}[\sigma(x_2 - x_3) - T_L] \end{cases}$$
(26)

Using Equations (25) and (26), the net control laws become

$$\begin{cases} u_{net_1} = x_1 - x_3 x_2 - U_d + \dot{d}_1 + 3x_1 + 3d_1 - 15\frac{d}{dt}(-x_1 + x_3 x_2 + U_d) - \zeta_1 sign(s_1) \\ u_{net_2} = x_1 + x_3 x_2 - \gamma x_3 - U_q + \dot{d}_2 - 5x_2 + 5d_2 - 18\frac{d}{dt}(-x_1 - x_3 x_2 + \gamma x_3 + U_q) - \zeta_2 sign(s_2) \\ u_{net_3} = -\sigma(x_2 - x_3) + T_L + \dot{d}_3 - 8x_3 + 5d_3 - 20\frac{d}{dt}[\sigma(x_2 - x_3) - T_L] - \zeta_3 sign(s_3) \end{cases}$$
(27)

Using the Lyapunov function candidate as $V(s) = \frac{1}{2}S_i^2$ and the assumption that $V(s)$ continuously differentiable function, the existence of stable sliding surfaces is ensured in the following manner. For surface s_1:

$$\dot{V}(s_1) = s_1 \dot{s}_1$$

$$\dot{V}(s_1) = s_1(\dot{e}_1 + 3e_1 + 15\frac{d^2}{dt}e_1)$$

$$\dot{V}(s_1) = s_1(-x_1 + x_3 x_2 + U_d - \dot{d}_1 + 3x_1 - 3d_1 + 15\frac{d}{dt}(-x_1 + x_3 x_2 + U_d) - 15\frac{d}{dt}(d_1) + u_{net_1})$$

Using the first Equation of u_{net_1} from Equation (27), we get $\dot{V}(s_1) = -s_1\zeta_1|sign(s_1)|$ where $\dot{V}(s_1) < 0$ since ζ_1 is positive from Barbelet lemma it is proved that the error $e(t) \to 0$ at $t \to \infty$. As a result, the stability condition is satisfied $\forall\ s$. Similarly, for surface s_2

$$\dot{V}(s_2) = s_2 \dot{s}_2$$

$$\dot{V}(s_2) = s_2(\dot{e}_2 + 5e_2 + 18\frac{d^2}{dt}e_2)$$

$$\dot{V}(s_2) = s_2[-x_1 - x_3 x_2 + \gamma x_3 + U_q - \dot{d}_2 + 5x_2 - 5d_2 +$$
$$+18\frac{d}{dt}(-x_1 - x_3 x_2 + \gamma x_3 + U_q) + 18\frac{d}{dt}(\dot{d}_1) + u_{net_2}]$$

By substituting the expression of u_{net_2} from equation (27), we get $\dot{V}(s_2) = -s_2\zeta_2|sign(s_2)|$. Since ζ_2 is positive so $\dot{V}(s_2) < 0$: negative definite and Barbelet lemma proves that the error $e(t) \to 0$ at $t \to \infty$.
Similarly, for surface s_3

$$\dot{V}(s_3) = s_3 \dot{s}_3$$

$$\dot{V}(s_3) = s_3(\dot{e}_3 + 5e_3 + 20\frac{d^2}{dt}e_3)$$

$$\dot{V}(s_3) = s_3(\sigma(x_2 + x_3) - T_L - \dot{d}_3 + 8x_3 - 5d_3 +$$
$$+20\frac{d}{dt}(\sigma(x_2 + x_3) + T_L) - 20\frac{d}{dt}(\dot{d}_1) - \zeta_3 sign(s_3)$$

$\dot{V}(s_3) = -s_3\zeta_3|sign(s_3)|$ by using the expression of u_{net_3} from Equation (27). where $\dot{V}(s_3) < 0$ and from Barbelet lemma it proves the error $e(t) \to 0$ at $t \to \infty$. As a result, the stability condition is satisfied $\forall\ s$.

3.4. Control of PMSM: Simulation Results

In this subsection, the performance of the PMSM system with P SMC, PI SMC, and PID SMC is demonstrated via MATLAB simulation. Reference values for all the three SMC techniques are considered as $d_1 = 17.21$, $d_2 = 1$, $d_3 = 0$. For simulation, the controller parameters are selected as:

- Parameters for P-SMC are $(n_1=2, n_2=0.8, n_3=0.8)$.
- Parameters for PI-SMC are $(n_1=2, n_2=0.8, n_3=0.8)$ and $(k_2=3, k_2=5, k_3=8)$.
- Parameters for PID-SMC are $(k_1=3, k_2=5, k_3=8)$ and $(D_1=15, D_2=18, D_3=20)$.

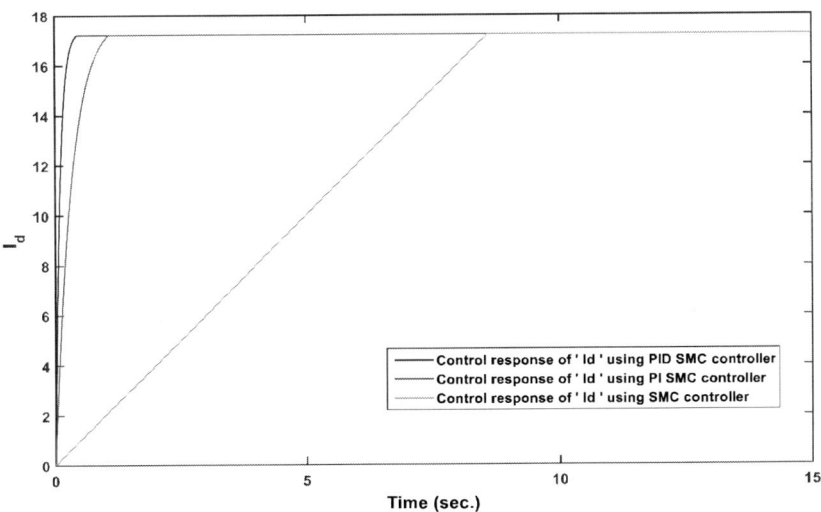

Figure 4. Controlled response of i_d using PID, PI, and P Sliding Mode Controllers.

Figure 4 shows the controlled response of direct axis current (i_d) of PMSM using P, PI, and PID SMC techniques. The P SMC technique controls the behaviour of direct axis current (i_d) at 8.8 seconds. PI SMC is showing controlled behaviour of direct axis current (i_d) at 1.1 seconds, and PID SMC shows controlled behaviour at 0.4 seconds. The controlled response of quadrature axis current (i_q) using the three controllers are shown in Figure 5. The P, PI, and PID SMC techniques are able to control the chaotic behaviour of quadrature axis current (i_q) in 1.2 seconds, 0.4 seconds, and 0.2 seconds, respectively.

Similarly, the speed control of permanent magnet synchronous motor using P, PI, and PID sliding mode control techniques is shown in Figure 6. The P, PI, and PID SMC techniques control the chaotic behaviours of speed at 0.0125, 0.012, and 0.0115 seconds.

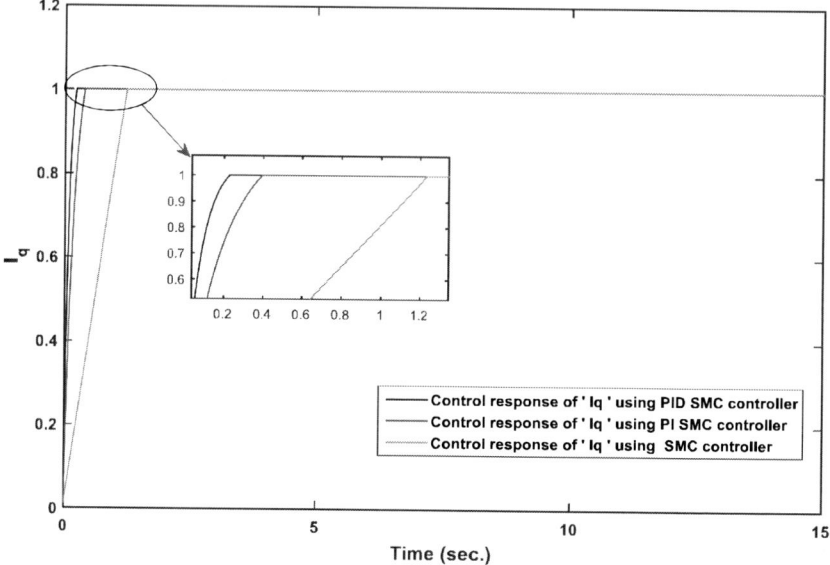

Figure 5. Controlled response of i_q using P, PI, and PID SMC techniques.

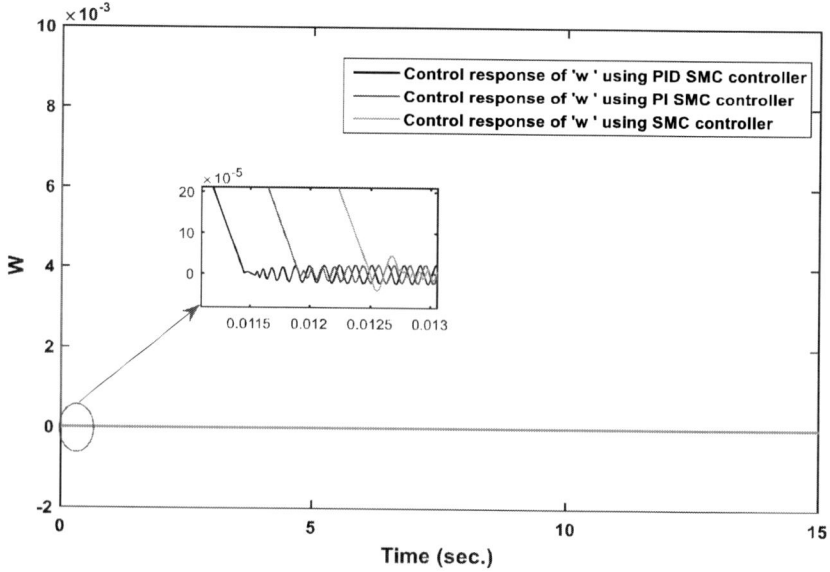

Figure 6. Controlled speed response (w) using PID, PI, and P sliding mode controls.

4. CASCADED PI SMC DESIGN AND STABILITY ANALYSIS FOR ELECTRIC VEHICLE

Proportional Integral (PI) control is proposed by using the same Proportional Integral (PI) block in cascade. In this cascaded PI controller, the error signal $e(t)$ is the input to the first PI controller and the output from the first controller $u'(t)$ is the input to the second PI controller. Therefore, the output of the second PI controller $u(t)$ will act as a controller to the PMSM plant to control the chaotic behaviour of the given reference signal $r(t)$.

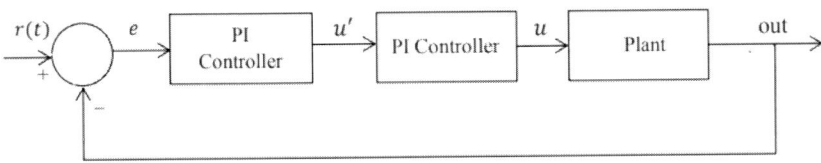

Figure 7. Block diagram of proposed cascaded PI technique.

Electric Vehicle (EV) dynamic model [13] is considered as:

$$\begin{cases} \frac{di}{dt} = \left(\frac{1}{L}\right)\{V - (R_a + R_f)i - L_{af}.i.\omega\} \\ \frac{d\omega}{dt} = \left(\frac{1}{j+m\left(\frac{r}{a}\right)^2}\right)\left\{L_{af}i^2 - B\omega - \left(\frac{r}{a}\right)\left(\mu_{rr}mg + \frac{1}{2\delta A c_d V^2} + mgsin\emptyset\right)\right\} \end{cases}$$

(28)

The state-space representation is obtained using the following assumptions.

$$\begin{cases} i = x_1, & \omega = x_2 \\ \frac{di}{dt} = \dot{x}_1, & \frac{d\omega}{dt} = \dot{x}_2 \end{cases}$$

EV dynamics in the state space form is written as:

$$\begin{cases} \dot{x}_1 = \left(\frac{1}{L}\right)\{V - (R_a + R_f)x_1 - L_{af}x_1x_2\} \\ \dot{x}_2 = \left(\frac{1}{j + m\left(\frac{r}{a}\right)^2}\right)\{L_{af}x_1{}^2 - Bx_2 - \left(\frac{r}{a}\right)\left(\mu_{rr}mg + \frac{1}{2\delta Ac_d V^2} + mgsin\emptyset\right)\} \end{cases}$$

(29)

Let the tracking error be defined as $e = x - d$ where $e \in R^2$. Then, the error for each state variable is defined as:

$$\begin{cases} e_1 = x_1 - d_1 \\ e_2 = x_2 - d_2 \end{cases} \quad (30)$$

Then the sliding surface

$$\begin{cases} s_1 = e_1 + (k_{PI})_1 \int_0^t e_1(\tau)d\tau + (k_I{}^*)_1 \int \frac{e_1(\tau)^2}{2} d\tau \\ s_2 = e_2 + (k_{PI})_2 \int_0^t e_2(\tau)d\tau + (k_I{}^*)_2 \int \frac{e_2(\tau)^2}{2} d\tau \end{cases} \quad (31)$$

where the value of PI constants is k_{PI}, $(k_I{}^*)_1$, $(k_{PI})_2$ and $(k_I{}^*)_2$ are 20, 100, 20 and 100, respectively. The sliding dynamics is written as:

$$\begin{cases} \dot{s}_1 = \dot{e}_1 + 20e_1 + 100\frac{e_1{}^2}{2} \\ \dot{s}_2 = \dot{e}_2 + 20e_2 + 100\frac{e_2{}^2}{2} \end{cases} \quad (32)$$

The system operates in the sliding mode when $\dot{s}_i = 0$, $i = 1, 2$. Therefore, the equivalent error dynamics is written as:

$$\begin{cases} \dot{e}_1 = \dot{x}_1 - \dot{d}_1 \\ \dot{e}_2 = \dot{x}_2 - \dot{d}_2 \end{cases} \quad (33)$$

To ensure the occurrence of sliding motion, the net control laws are defined as:

$$\begin{cases} u_{net1} = U_{eq1} - \zeta_1 sign(s_1) \\ u_{net2} = U_{eq2} - \zeta_2 sign(s_2) \end{cases} \quad (34)$$

where ζ_1 and ζ_1 are positive constants. The equivalent control lows in defined as:

$$\begin{cases} U_{eq1} = (-\frac{V}{L} + (\frac{1}{L})x_1 x_2 - (k_{PI})_1 x_1 + (k_{PI})_1 d_1 - \frac{(k_I)^*}{2}(x_1 - d_1)^2) \\ U_{eq2} = \left(\frac{1}{j+m(\frac{r}{a})^2}\right)\{-L_{af} x_1^2 + (\frac{r}{a})(\mu_{rr} mg + \frac{1}{2\delta A c_d V^2} + mg sin\emptyset)\} \\ \qquad -(k_{PI})_1 x_1 + (k_{PI})_1 d_1 - \frac{(k_I)^*}{2}(x_1 - d_1)^2) \end{cases} \quad (35)$$

Then net control becomes

$$\begin{cases} u_{net1} = \left(-\frac{V}{L} + (\frac{1}{L})x_1 x_2 - (k_{PI})_1 x_1 + (k_{PI})_1 d_1 - \frac{(k_I)^*}{2}(e_1)^2\right) - \zeta_1 sign(s_1) \\ u_{net2} = \left(\frac{1}{j+m(\frac{r}{a})^2}\right)\{-L_{af} x_1^2 + (\frac{r}{a})(\mu_{rr} mg + \frac{1}{2\delta A c_d V^2} + mg sin\emptyset)\} - -(k_{PI})_1 x_1 \\ \qquad +(k_{PI})_1 d_1 - \frac{(k_I)^*}{2}(e_2)^2) - \zeta_2 sign(s_2) \end{cases} \quad (36)$$

The general expression of PI controller is defined as $u_i = k_p' e_i + k_I' \int_0^t e_i(\tau) d\tau$ for $i = 1, 2$. Therefore, the first PI controller will have the following expression

$$u' = k_p' e + k_I' \int_0^t e_i(\tau) d\tau \quad (37)$$

Let the second PI controller be defined as:

$$u = k_p(u') + k_I \int_0^t u'(\tau) d\tau \quad (38)$$

Now by using the u' expression in Equation (38), we can derive cascaded PI control law as:

$$u = k_p[k_p'e + k_I' \int_0^t e(\tau)d\tau + k_I \int_0^t [k_p'e + k_I' \int_0^t e(\tau)d\tau]$$
$$u = k_p k_p' e + k_p k_I' \int_0^t e(\tau)d\tau +$$
$$+ k_I k_p' \int_0^t e(\tau)d\tau + k_I k_I' \iint_0^t e(\tau)d\tau]$$
$$u = k_p k_p' e + [k_p k_I' + k_I k_p'] \int_0^t e(\tau)d\tau + k_I k_I' \int_0^t \frac{e(\tau)^2 d\tau}{2}$$
$$u = k_p' e + k_{PI} \int_0^t e(\tau)d\tau + k_I' \int_0^t \frac{e(\tau)^2 d\tau}{2} \qquad (39)$$

where the gains of cascaded PI control law are $k_p = k_p k_p'$, $k_{PI} = k_p k_I' + k_I k_p'$, $k_I^* = k_I k_I'$.

Let the Lyapunov function candidate be defined as $V(s) = \frac{1}{2}s_i^2$ for $i = 1, 2$. To ensure the existence of a stable sliding surface s_1 assuming $V(s)$ the continuously differentiable function then

$$\dot{V}(s_1) = s_1 \dot{s}_1$$
$$\dot{V}(s_1) = s_1(\dot{e}_1 + 20e_1 + 50e_1^2)$$
$$\dot{V}(s_1) = s_1[(\dot{x}_1 - d_1) + 20(x_1 - d_1)50(x_1 - d_1)^2]$$
$$= s_1 \left[\frac{1}{L}\{V - (R_a + R_f)x_1 - L_{af}x_1x_2\} - d_1 20x_1 - 20d_1 + +50x_1^2 + 50d_1^2 - 100x_1 d_2\right] + u_{net1} \qquad (40)$$

Using Equation (40) and by substituting the value of u_{net1} from Equation (36), we get

$$\dot{V}(s_1) = s_1(-\zeta_1 sign(s_1))$$
$$\dot{V}(s_1) = -s_1\zeta_1 |sign(s_1)|$$

$\dot{V}(s_1) < 0$: negative definite for the positive value of ζ_1. From Barbelet lemma, it proves the $e(t) \to 0$ at $t \to \infty$, and the sliding surface is asymptotically stable $\forall$ S.

Similarly, for surface s_2

$$\dot{V}(s_2) = s_2\dot{s}_2$$
$$\dot{V}(s_1) = s_1(\dot{e}_1 + 20e_2 + 50e_2{}^2)$$
$$\dot{V}(s_2) = s_2[(\dot{x}_2 - \dot{d}_2) + 20(x_2 - d_2) + 50(x_2 - d_2)^2]$$
$$= s_2[\dot{x}_2 - \dot{d}_2 + 20x_1 - 20d_1 + 50x_1{}^2 + 50d_1{}^2 - 100x_1 d_2]$$
$$= s_2\left[\left(\frac{1}{j+m\left(\frac{r}{a}\right)^2}\right)\left\{-L_{af}x_1{}^2 - Bx_2 - \left(\frac{r}{a}\right)\left(\mu_{rr}mg + \frac{1}{2\delta A c_d V^2} + mgsin\emptyset\right)\right\} - \dot{d}_2 + 20x_1 - 20d_1 + 50x_1{}^2 + 50d_1{}^2 - 100x_1 d_2\right] +$$
$$u_{net2} \tag{41}$$

Using Equation (41) and by substituting the expression of u_{net2}, we get $\dot{V}(s_2) = s_2\bigl(-\zeta_2 sign(s_2)\bigr)$ as:

$$\dot{V}(s_2) = -s_2\zeta_2 |sign(s_2)|$$

$\dot{V}(s_2)$ is negative definite for $\zeta_2 > 0$. Therefore, from Barbelet lemma $e(t) \to 0$ at $t \to \infty$. The result of cascaded PI sliding mode control is discussed in the next section.

5. DESIGN AND STABILITY OF PID SMC FOR ELECTRIC VEHICLE

Assuming the tracking error as defined in (30), the PID sliding surfaces are defined as

$$\begin{cases} s_1 = e_1 + \int_0^t k_1 e_1(\tau)d\tau + D_1 \frac{d}{dt}e_1 \\ s_2 = e_2 + \int_0^t k_2 e_2(\tau)d\tau + D_2 \frac{d}{dt}e_2 \end{cases} \tag{42}$$

where integral gains $k_1 = k_2 = 6$ and the differential gains D_1, D_2 are 12 and 10, respectively. Then, the sliding dynamics is written as:

$$\begin{cases} \dot{s}_1 = \dot{e}_1 + 6e_1 + 12\dfrac{d^2}{dt^2}e_1 \\ \dot{s}_2 = \dot{e}_2 + 6e_2 + 10\dfrac{d^2}{dt^2}e_2 \end{cases} \quad (43)$$

The system operates in the sliding mode when $s_i = 0$. To ensure the occurrence of sliding motion, the net control laws are defined as

$$\begin{cases} u_{net_1} = u_{eq_1} - \zeta_1 sign(s_1) \\ u_{net_2} = u_{eq_2} - \zeta_2 sign(s_2) \end{cases} \quad (44)$$

The equivalent control laws are defined as:

$$\begin{cases} u_{eq_1} = -\left[\left(\dfrac{1}{L}\right)\{V - (R_a + R_f)x_1 - L_{af}x_1x_2\} - d_1\right] - 6e_1 - 12\dfrac{d}{dt}\left[\left(\dfrac{1}{L}\right)\{V - (R_a + R_f)x_1 - L_{af}x_1x_2\} - d_1\right] \\ u_{eq_2} = -\left[\left(\dfrac{1}{j+m\left(\frac{r}{a}\right)^2}\right)\{L_{af}x_1{}^2 - Bx_2 - \left(\dfrac{r}{a}\right)\left(\mu_{rr}mg + \dfrac{1}{2\delta A c_d V^2} + mg\sin\emptyset\right)\} - d_2\right] - 6e_2 \\ \qquad -10\dfrac{d}{dt}\left[\left(\dfrac{1}{j+m\left(\frac{r}{a}\right)^2}\right)\{L_{af}x_1{}^2 - Bx_2 - \left(\dfrac{r}{a}\right)\left(\mu_{rr}mg + \dfrac{1}{2\delta A c_d V^2} + mg\sin\emptyset\right)\} - d_2\right] \end{cases}$$

(45)

The net control laws from Equation (44) becomes

$$\begin{cases} u_{net_1} = -\left[\left(\dfrac{1}{L}\right)\{V - (R_a + R_f)x_1 - L_{af}x_1x_2\} - d_1\right] - 6e_1 - 12\dfrac{d}{dt}\left[\left(\dfrac{1}{L}\right)\{V - (R_a + R_f)x_1 - L_{af}x_1x_2\} - d_1\right] - \zeta_1 sign(s_1) \\ u_{net_2} = -\left[\left(\dfrac{1}{j+m\left(\frac{r}{a}\right)^2}\right)\{L_{af}x_1{}^2 - Bx_2 - \left(\dfrac{r}{a}\right)\left(\mu_{rr}mg + \dfrac{1}{2\delta A c_d V^2} + mg\sin\emptyset\right)\} - d_2\right] - 6e_2 \\ \qquad -10\dfrac{d}{dt}\left[\left(\dfrac{1}{j+m\left(\frac{r}{a}\right)^2}\right)\{L_{af}x_1{}^2 - Bx_2 - \left(\dfrac{r}{a}\right)\left(\mu_{rr}mg + \dfrac{1}{2\delta A c_d V^2} + mg\sin\emptyset\right)\} - d_2\right] - \zeta_2 sign(s_2) \end{cases}$$

(46)

Let the Lyapunov function candidate be defined as $V(s) = \dfrac{1}{2}s_i^2$. Assuming $V(s)$ continuously differentiable function to ensure a stable sliding surface.

$$\dot{V}(s_1) = s_1\dot{s}_1$$

$$\dot{V}(s_1) = s_1(\dot{e}_1 + 6e_1 + 12\frac{d^2}{dt}e_1)$$

$$\dot{V}(s_1) = s_1[\left(\frac{1}{L}\right)\{V - (R_a + R_f)x_1 - L_{af}x_1x_2\} - \dot{d}_1 + 6(\text{ }_1 - d_1) + 12\frac{d}{dt}\left[\left(\frac{1}{L}\right)\{V - (R_a + R_f)x_1 - L_{af}x_1x_2\} - \dot{d}_1\right] + u_{net_1}]$$

Using u_{net_1} from Equation (46), we get $\dot{V}(s_1) = -s_1\zeta_1|sign(s_1)|$.

$\dot{V}(s_1) < 0$ since ζ_1 is positive. Therefore, Barbelet lemma proves that the error $e(t) \to 0$ at $t \to \infty$. As a result, the stability condition is satisfied $\forall$ s.

Similarly, for surface s_2

$$\dot{V}(s_2) = s_2\dot{s}_2$$

$$\dot{V}(s_2) = s_2(\dot{e}_2 + 6e_2 + 10\frac{d^2}{dt}e_2)$$

$$\dot{V}(s_2) = s_2\left[\left(\frac{1}{j+m\left(\frac{r}{a}\right)^2}\right)\{L_{af}x_1^2 - Bx_2 - \left(\frac{r}{a}\right)\left(\mu_{rr}mg + \frac{1}{2\delta A c_d V^2} + mgsin\emptyset\right)\} - \dot{d}_2 + 6(x_2 - d_2) + 10\frac{d}{dt}\left[\left(\frac{1}{j+m\left(\frac{r}{a}\right)^2}\right)\{L_{af}x_1^2 - Bx_2 - \left(\frac{r}{a}\right)\left(\mu_{rr}mg + \frac{1}{2\delta A c_d V^2} + mgsin\emptyset\right)\} - \dot{d}_2\right] + u_{net_2}\right]$$

By substituting the value of u_{net_2} from Equation (46), $\dot{V}(s_2)$ is obtained as

$$\dot{V}(s_2) = -s_2\zeta_2|sign(s_2)|$$

Since $\dot{V}(s_2) < 0$ for $\zeta_2 > 0$ and Barbelet lemma proves that the error $e(t) \to 0$ at $t \to \infty$. As a result, the stability condition is satisfied $\forall$ s.

6. SPEED CONTROL OF ELECTRIC VEHICLE: SIMULATION RESULTS

EV performance with cascaded PI SMC and PID SMC is simulated in a MATLAB environment. The controller parameters are selected as:

- Parameters for Cascaded PI SMC are ($d_1 = 0.5$, $d_2 = 0.2$), ($n_1 = 2.5$, $n_2 = 1.2$) and ($k_1 = k_2 = 20$, $K_1 = K_2 = 100$).
- Parameters for PID SMC are ($d_1 = 0.5$, $d_2 = 0.2$), ($n_1 = 2.5$, $n_2 = 1.2$), ($k_1 = k_2 = 6$) and ($D_1 = 12$, $D_2 = 10$).

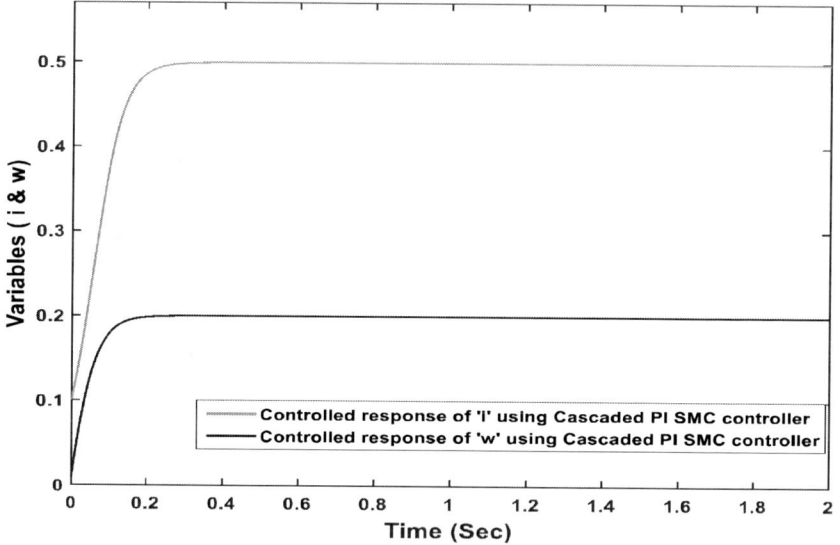

Figure 8. Controlled time response of 'i' and 'ω' using Cascaded PI SMC.

Figure 8 shows the controlled response of current (i) and speed (ω) of Electric Vehicle using cascaded PI SMC techniques. Cascaded PI sliding mode control can control the current (i) in 0.38 seconds and speed (ω) in 0.24 seconds. The required control efforts are shown in Figure 9, showing the control effort required to control the current (i) and speed (ω).

Chaos Control in Electric Vehicle using Cascaded PI and PID ...

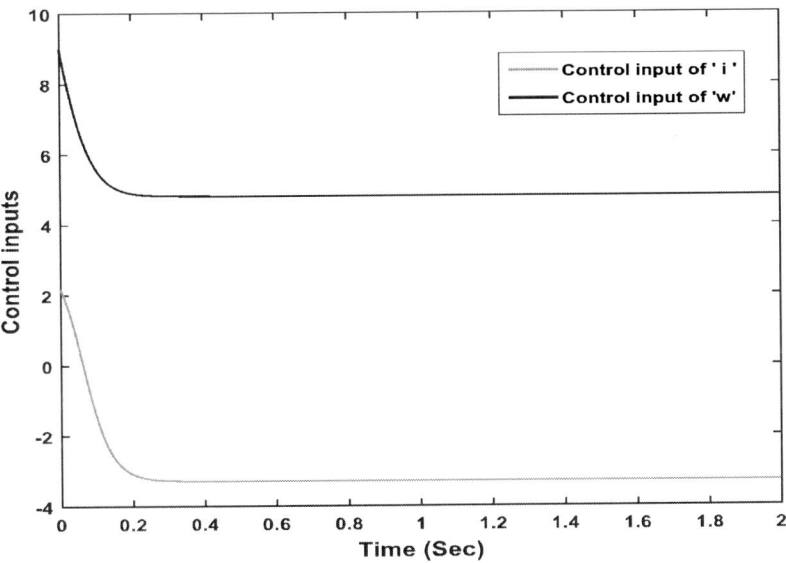

Figure 9. Required control inputs in cascaded PI SMC for 'i' and 'ω' variables.

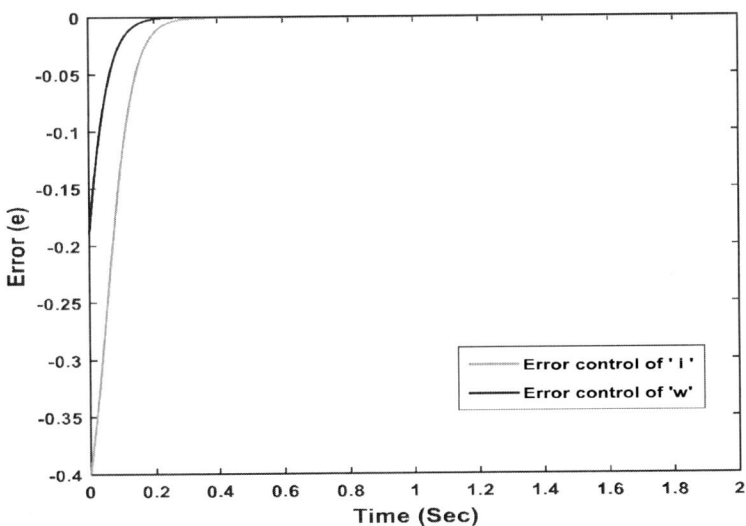

Figure 10. Tracking error for current 'i' and speed' ω'.

The tracking error for current (i) and speed (ω) is shown in Figure 10. The tracking error of EV current (i) goes to zero at 0.38 seconds, and the tracking error of speed (ω) goes to zero at 0.24.

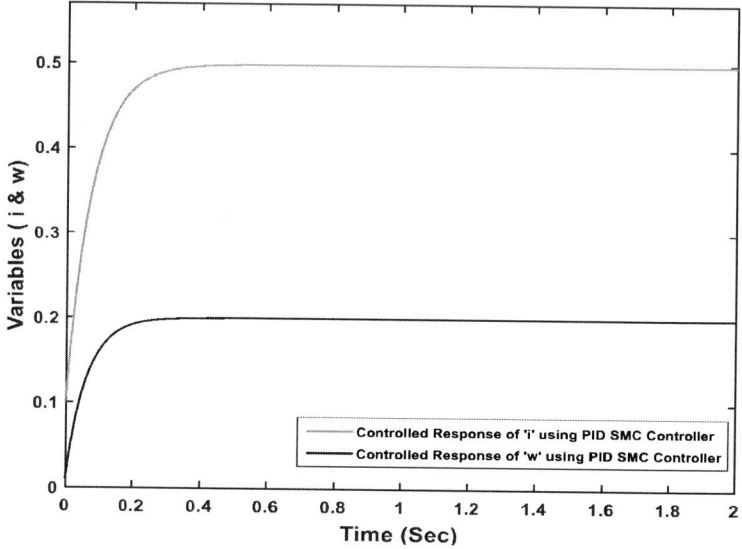

Figure 11. Controlled states 'i' and 'ω' using PID SMC.

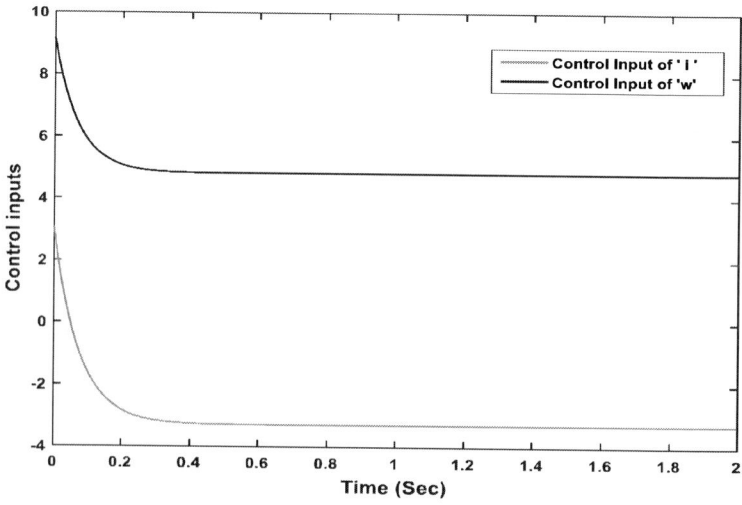

Figure 12. PID sliding mode control inputs for current 'i' and speed 'ω'.

Figure 11 shows the controlled response of current (i) and speed (ω) variables of EV using PID SMC technique, and it is apparent that current (i) is being controlled in 0.60 seconds, and speed (ω) is being controlled to the reference point in 0.38 seconds. The required control effort to control the EV is shown in Figure 12.

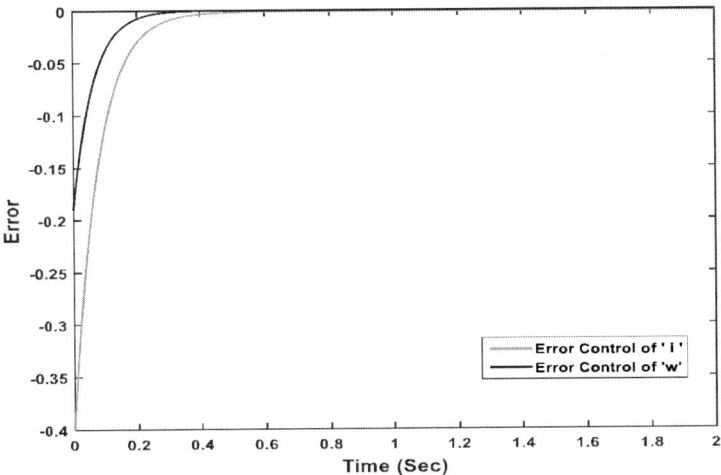

Figure 13. Tracking error of states in PID SMC technique.

Figure 13 shows the tracking errors for current (i) and speed (ω) variables of EV in case of PID sliding mode control technique which confirms that the tracking error for current (i) goes to zero in 0.57 seconds and for speed (ω) of EV, it goes to zero in 0.35 seconds.

Figure 14 shows the controlled response of EV current (i) using both cascaded PI and PID SMC techniques. Cascaded PI SMC has a settling time of 0.38 seconds, and PID SMC has 0.60 seconds. The current control of EV using cascaded PI sliding mode control is better than the PID SMC technique. Similarly, In Figure 15, the speed (ω) controlled response of electric vehicles where the cascaded PI SMC has a settling time of 0.24 second PID SMC has a settling time of 0.38 seconds. The speed control using cascaded PI sliding mode control is better than the PID SMC technique.

7. COMPARATIVE PERFORMANCE ANALYSIS AND DISCUSSION

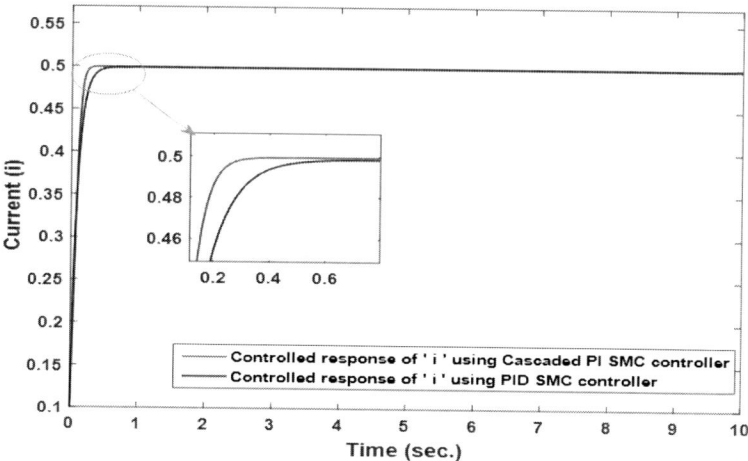

Figure 14. Controlled current 'i' response using cascaded PI and PID SMC technique.

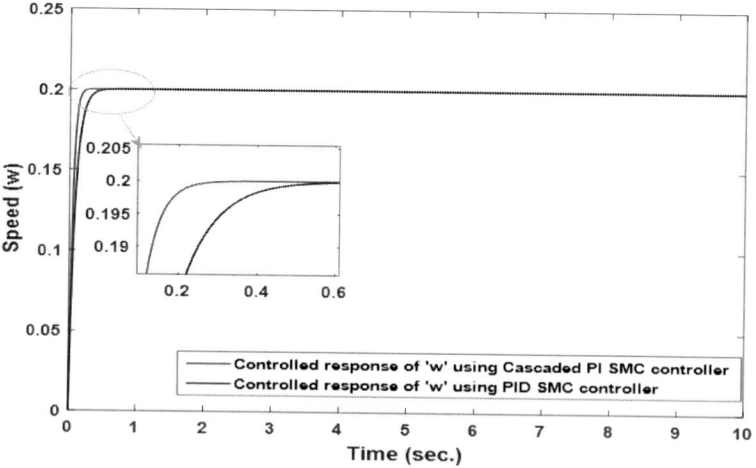

Figure 15. Controlled speed 'ω' response using cascaded PI SMC and PID SMC techniques.

Further, the average error is calculated as $e_{avg} = \frac{1}{2}[e_1^2 + e_2^2]$ to have more concrete evidence on the performance of both the controllers. The average error is plotted for cascaded PI and PID SMC techniques and shown in Figures 16 and 17, respectively.

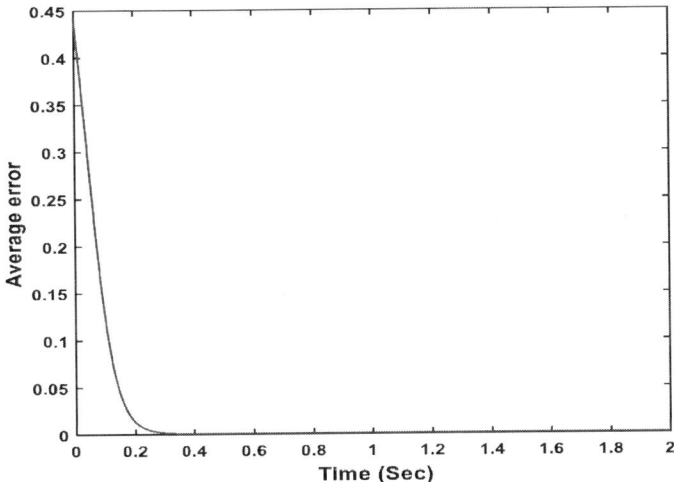

Figure 16. The average error in the cascaded PI sliding mode control of the electric vehicle.

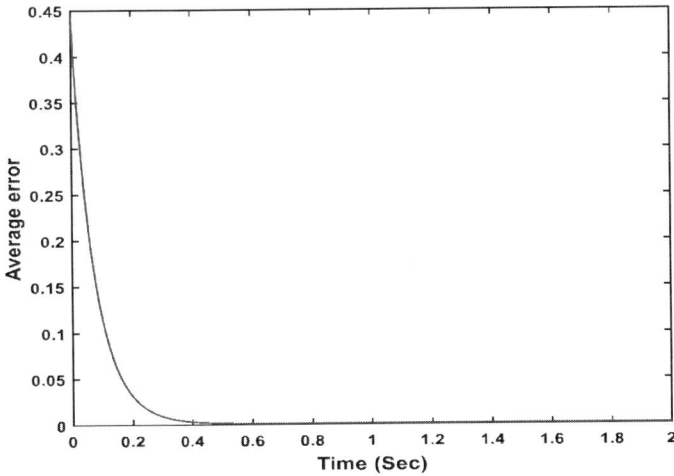

Figure 17. The average error in the PID SMC of the electric vehicle.

Figure 16 shows the average error in the electric vehicle's cascaded PI sliding mode control. The average tracking error goes to zero at 0.28

seconds. The average error in PID SMC of an electric vehicle is shown in Figure 17, which confirms that the average error goes to zero in 0.45 seconds. It is concluded that in the control of EV, the performance of cascaded PI sliding mode control is better than the PID SMC technique.

CONCLUSION AND FUTURE SCOPE

In this chapter, three control techniques, classical P-SMC, PI–SMC, PID–SMC are discussed to suppress the chaos in PMSM motor and two control techniques; PID–SMC and Cascaded PI–SMC, to control the speed and current of EV. The comparative analysis is performed by keeping the same parameter values in different controllers that control the PMSM and EV. The disadvantage of the classical sliding mode control is the phenomenon of "chattering". We have developed PI-SMC, PID-SMC, and Cascaded PI–SMC techniques to minimise the chattering phenomenon. All the simulation is performed in a MATLAB environment. The simulation results reveal that in the speed control of PMSM using P-SMC, PI-SMC, PID-SMC techniques, the performance order is $T_{PID-SMC} < T_{PI-SMC} < T_{P-SMC}$ in terms of settling time and chattering order is (PID − SMC < PI − SMC < P − SMC) This ensures that the PID sliding mode control has better control performance in terms of chattering and tracking performance. In the case of speed control of Electric Vehicle using PID-SMC and Cascaded PI–SMC technique, the settling time $T_{Cas.PI-SMC} < T_{PID-SMC}$ and chattering order (Cascaded PI − SMC < PID − SMC). This ensures that the cascaded PI sliding mode control has better control performance in terms of chattering and tracking performance.

REFERENCES

[1] Su Y. X., C. H. Zheng, and B. Y. Duan, Automatic disturbances rejection controller for precise motion control of permanent-magnet synchronous motors, *IEEE Trans. Ind. Electron.*, 52, 814-823, (2005).

[2] Li S. and Z. Liu, Adaptive speed control for permanent magnet synchronous motor system with variations of load inertia, *IEEE Trans. Ind. Electron.*, 56, 3050-3059, (2009).

[3] Wang L., T. Chai, and L. Zhai, Neural-network-based terminal sliding mode control of robotic manipulators including actuator dynamics, *IEEE Trans. Ind. Electron.*, 56, 3296-3304, (2009).

[4] Chiu J. -C., K.-S. Leung, and H.-H. Chung, High-order switching surface in boundary control of inverters, *IEEE Trans. Power Electron.*, 22, 1753-1765, (2007).

[5] Fnaiech M. A., F. Betin, G.-A. Capolino, and F. Fnaiech, Fuzzy logic and sliding-mode controls applied to six-phase induction machine with open phases, *IEEE Trans. Ind. Electron.*, 57, 354-364, (2010).

[6] Feng Y., J. F. Zheng, X. H. Yu, and N. Vu Truong, Hybrid terminal sliding mode observer design method for a permanent magnet synchronous motor control system, *IEEE Trans. Ind. Electron.*, 56, 3424-3431, (2009).

[7] Castillo-Toledo B., S. Di Gennaro, A. G. Loukianov, and J. Rivera, Hybrid control of induction motors via sampled closed representations, *IEEE Trans. Ind. Electron.*, 55, 3758-3771, (2008).

[8] Lin F. J., J. C. Hwang, P. H. Chou, and Y. C. Hung, FPGA-based intelligent-complementary sliding-mode control for PMSM servo-drive system, *IEEE Trans. Power Electron.*, 25, 2573-2587, (2010).

[9] Arrillaga J., B. C. Smith, N. R. Watson, A. R. Wood, *Power System Harmonic Analysis*, John Wiley & Sons Ltd., (2003).

[10] Lin F. J., Y. C. Hung, J. C. Hwang, I. P. Chang, M. T. Tsai, Digital signal processor-based probabilistic fuzzy neural network control of in-wheel motor drive for light electric vehicle, *IET Elect. Power Appl.*, 6, 47–61, (2012).

[11] Mura R., V. Utkin, S. Onori, Energy management design in hybrid Electric vehicles: a novel optimality and stability framework, *IEEE Trans. Control System Technol.*, 23, 1307-1322, (2015).

[12] Overington S., S. Rajakaruna, High efficiency control of internal combustion engines in blended charge depletion/charge sustenance strategies for plug-in hybrid electric vehicles, *IEEE Trans. Veh. Technol.*, 64, 48-61, (2015).

[13] Huang Q., Z. Huang, H. Zhou, Nonlinear optimal and robust speed control for a light-weighted all-electric vehicle, *IET Control Theory Application*, 3, 437-444, (2009).

[14] Huang Q., J. Li, Y. Chen, *Control of electric vehicle*, Intech Open Publisher, (2010).

[15] Iqbal, G. K. Singh, Chaos control of permanent magnet synchronous motor using simple controllers, *Trans. Inst. Meas. Control*, 41, 2352–2364, (2019).

In: Nonlinear Systems
Editor: P. P. Singh
ISBN: 978-1-68507-660-3
© 2022 Nova Science Publishers, Inc.

Chapter 4

CHAOS CONTROL IN POWER SYSTEM USING HIGHER ORDER SLIDING MODE CONTROL

Manish Kumar[*] and Piyush Pratap Singh[†]

Department of Electrical Engineering, National Institute
of Technology Meghalaya, Shillong, India

Abstract

This chapter puts forward the existence of chaos and its control in the extended single machine infinite bus (SMIB) power system using the higher order sliding mode control (HOSMC) technique. The conventional SMIB power system is extended by considering the generator's transient and excitation voltages. Chaos behaviour is observed and validated using a phase portrait plot and the calculation of the Lyapunov exponent. A HOSMC technique is used to suppress the chaos in the extended four-dimensional SMIB power system model. Second-order sliding surfaces are designed using pole placement to ensure the stability of error dynamics. Modified sliding mode control laws are designed to provide the reachability of the designed sliding surfaces. The required stability and reachability conditions are derived using Lyapunov stability theory. The HOSMC technique is

---

[*] Corresponding Author's E-mail: 58manish.sit@gmail.com.
[†] Corresponding Author's E-mail: piyushpratapsingh@gmail.com.

compared with conventional SMC and proportional-integral SMC techniques. Two measures, the average error energy and average control energy, are considered. But due to trade-offs between the measures, a performance comparison is achieved using three feasible performance indices. Simulation is done in the MATLAB Simulink environment. Numerical simulation results validate the theoretical approach and reveal the achievement of the objectives.

1. INTRODUCTION

Chaos has been investigated in many nonlinear dynamic systems, such as nonlinear circuits, permanent magnet synchronous motors, power transformers and power systems [1-10], etc. The power system is a highly nonlinear dynamic system. When a power system operates near the stability zone, a sudden disturbance or parameter variations may cause chaotic oscillation or even blackout [2]. The control of such chaotic oscillations is required and becomes the primary motivation of study.

A passive controller is first presented in [3], transforming the chaotic single machine infinite bus (SMIB) power system into an equivalent passive system. It is proved that the equivalent system can be asymptotically stabilised at any desired fixed point. A complex nonlinear phenomenon in the SMIB power system taking excitation hard limits into account is presented [4]. A global state feedback linearisation control technique is applied to the power system to control the chaotic behaviour in [5]. Linear and nonlinear state feedback controllers are proposed to control the bifurcation in the power system [6]. The behaviour of duffing oscillator revealing a wide range of phenomena specific to nonlinear systems is reported in [7]. Using fractional-order backstepping control, an adaptive controller is proposed to suppress chaos in the fractional-order SMIB power system [8]. A linear quadratic regulator (LQR) is designed to suppress the chaotic behaviour of PMSM. This is the first time when the robustness of LQR is used for chaos suppression of PMSM. Due to its robustness, the proposed LQR scheme eliminates the chaos and provides better performance than the conventional method [9]. Chaos synchronisation between unidirectional coupled ordinary differential and delay differential equations-based power system is investigated [10].

The adaptive network-based fuzzy inference system-based composite controller (ANFIS-CC) is proposed to resolve chaotic oscillations' instability and voltage collapse [11]. An adaptive control law controls chaotic oscillations in a simple interconnected power system [12]. A novel control scheme is proposed to achieve fast fixed-time system stabilisation based on fixed-time stability theory [11]. The ANFIS-CC, a more efficient computation scheme than a Mamdani fuzzy logic controller, to resolve instability problem is reported in [13]. In paper [14], bifurcations are observed for a power system model and a nonlinear controller is used to control the subcritical Hopf and period-doubling bifurcations to mitigate voltage collapse. The complex nonlinear dynamic phenomenon and its control in the SMIB power system using quasi-sliding mode control are investigated in [15]. Chaos synchronisation at master-slave philosophy is reported [16].

Chaos is suppressed for a four-dimensional fundamental power system model using a novel adaptive feedback controller [17]. A decentralised algorithm for real-time control of oscillatory dynamics in power systems with an extended linear quadratic regulator for optimal control is reported in [18]. In paper [19], LQR design to suppress the chaotic behaviours of PMSM is investigated. An adaptive controller based on a fractional-order backstepping method is proposed to suppress chaos in the FOSMIB power system is reported in [20]. An extended linear quadratic regulator and fast fixed-time non-singular sliding mode control to achieve the power system stabilisation are reported in [21] and [22], respectively. The fundamentals of forced oscillation in multimachine power systems are investigated by analysing and detecting forced oscillation to avoid confusion with modal oscillation [23]. Chaos control, synchronisation, and stability analysis of Vallis and El-Nino systems with fractional order are studied in [24]. Nonlinear control for decentralised control of power system dynamics is proposed [25]. An outlook on applying adaptive model predictive control for uncertain systems to stimulate new research interests for wind turbine systems is provided in [26]. A digital chaotic system is proposed based on the chaos generation strategy controlled by random sequences, and FPGA circuit implementation is reported [27].

However, the sliding-mode control (SMC) techniques are used for order reduction, decoupling design procedure, disturbance rejection and insensitivity to parameter variations. The control algorithm uses a variable structure system [28]. A SMC technique with a proportional integral (PI) switching surface is designed to control the position trajectory and reduce the undesirable chattering phenomenon [29]. Chattering occurs in conventional SMC, and it can be reduced by replacing the discontinuous sign function with a proper continuous function. A quasi-SMC is proposed based on delay function to control the chaotic oscillation [15]. Composite nonlinear feedback control and proportional integral sliding mode control (PISMC) is designed to reduce overshoot and fast response [30]. In paper [31], a PISMC technique reduces the chattering phenomenon. The comparative performance of synchronisation [32] and anti-synchronisation [33], between different classes of chaotic systems using nonlinear active, PISMC and adaptive control techniques. In paper [34], a SMC regulates frequency error and tie-line power error of a three-area interconnected power system. Chaos control of four-dimensional fundamental power system using pole placement based PISMC is reported in [35]. An integral terminal SMC design is proposed to enhance the power quality of wind turbines under unbalanced voltage conditions is presented in [36]. This paper assesses the proposed design under severe voltage sag conditions and parameter variations.

Literature reveals that most of the reported works are based on the simple SBIM power system, where transient voltage and the generator's excitation are not considered. It is well known that the effect of these variables introduces nonlinearity in the power system, which may bring the system into an unstable mode. Therefore, this work is motivated by the above issues, control of the SMIB power system, under the influence of transient voltage and the excitation voltage, using higher order sliding mode (HOSMC). The contributions of this chapter are:

1. To explore the SMIB power system under transient and excitation voltage influence.

2. To design the pole placement based on HOSMC law to suppress the chaotic oscillation present in the extended SMIB power system.

The existence of chaos in the extended SMIB power system and its control using the HOSMC technique is not available in the literature and reflects the novelty of this work.

The rest of the chapter is organised as follows. In Section 2, the description and chaotic behaviour of the extended SMIB power system are presented, and the problem is formulated. A modified higher-order sliding mode control design with relay characteristics is presented in Section 3. Section 4 contains simulation results and a comparative study. Finally, concluding remarks and future scopes are given in Section 5.

2. POWER SYSTEM DESCRIPTION AND PROBLEM FORMULATION

This section gives a brief description of the extended SMIB power system model. The simple SMIB power system model is shown in Figure 1, which is being used for stability analysis in power systems [37].

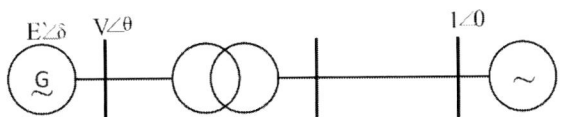

Figure 1. Single machine infinite bus (SMIB) power system.

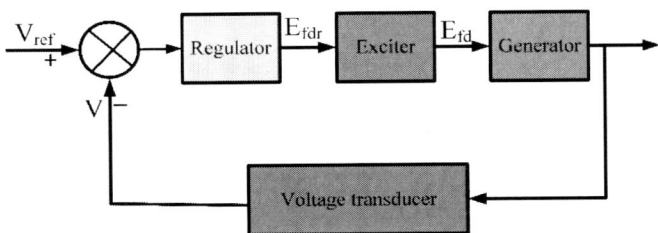

Figure 2. Block diagram representation of generator exciter system.

The dynamics (1) of the fundamental power system model was proposed by Ji and Venkatasubramanian [4].

The dynamic model of synchronous generator is expressed as:

$$\begin{cases} \dot{\delta} = 2\pi f_0 \omega \\ M\dot{\omega} = -D\omega + P_m - P_G \\ T'_{d0}\dot{E}' = -\frac{x_d + x}{x'_d + x} E' + \frac{x_d - x'_d}{x'_d + x} \cos\delta + E_{fd} \end{cases} \quad (1)$$

where parameters δ, ω and E' are generator angle, frequency deviation and transient voltage, respectively. M and D denote rotational inertia and damping coefficient of the generator, respectively. x_d, x'_d and x refer to d-axis synchronous reactance, d-axis transient reactance and total reactance of transformer and transmission line. Parameter f_0 is standard frequency of power system. T'_{d0} represents d-axis transient time-constant. E_{fd} and P_m stand for field voltage and the mechanical power of the equivalent generator, respectively. P_G is referred to as generated power which is expressed as:

$$P_G = \frac{E'}{x'_d + x} \sin\delta \quad (2)$$

Equation (1) contains the swing equation, the angle dynamics of the generator and model the of electromagnetics without damper windings. The SMIB power system model on excitation regulator dynamics and typical excitation system for the synchronous generator is shown in Figure 2.

The terminal voltage transducer senses generator terminal voltage (V) and compares it with the reference voltage (V_{ref}). An error signal is generated, which acts as an input to the excitation regulator. The regulator processes and amplifies the input signal and forms exciter control input (E_{fdr}). The exciter control input passes through an exciter and automatic voltage regulator and forms the field voltage (E_{fd}). The excitation regulator dynamics is described in (3).

$$\dot{E}_{fdr} = -K_A(V - V_{ref}) - E_{fdr} \qquad (3)$$

where E_{fdr} is field control signal, T_A is the excitation control time constant. K_A refers to the control gain, V_{ref} is the reference point of the generator terminal voltage (V). The terminal voltage of the generator is represented as:

$$V = \frac{1}{x+x'_d}\sqrt{(x'_d + xE'\cos\delta)^2 + (xE'\sin\delta)^2} \qquad (4)$$

Assuming the dynamics of the excitation regulator is insignificant for system dynamics. The output of the automatic voltage regulator (AVR) is considered as follows:

$$\begin{cases} E_{fd\,max} \text{ if } E_{fdr} > E_{fd\,max} \\ E_{fd} = E_{fd\,r} \text{ if } E_{fd\,min} \leq E_{fdr} \leq E_{fd\,max} \\ E_{fd\,min} \text{ if } E_{fdr} < E_{fd\,min} \end{cases} \qquad (5)$$

where E_{fdmin} and E_{fdmax} are upper and lower bounds, respectively, of the original signal E_{fdr}. Therefore, combining (3) to the conventional SMIB power system (1) results in a four-dimensional dynamic of the SMIB power system as:

$$\begin{cases} \dot{\delta} = 2\pi f_0 \omega \\ \dot{\omega} = \frac{1}{M}(-D\omega + P_m - P_G) \\ \dot{E}' = \frac{1}{T'_{do}}\left(-\frac{x_d+x}{x'_d+x}E' + \frac{x_d-x'_d}{x'_d+x}\right) \\ \dot{E}_{fdr} = \frac{1}{T_A}(-K_A(V - V_{ref}) - E_{fdr}) \end{cases} \qquad (6)$$

It is observed that the extended SMIB power system follows the route to chaos by varying the damping parameter (D). Variation of the parameter ($D = 19.8$ to $D = 5.0$) shows the period-doubling bifurcation. As the parameter goes to $D = 2.0$, it shows chaotic nature. System (6) shows chaotic behaviour for parameters $f_0 = 60$, $M = 10$, $P_m = 1.3$, $x'_d =$

$0.4, x = 0.5, T'_{d0} = 10, x_d = 1, T_A = 1, V_{ref} = 1.05, T'_{d0} = 10, x_d = 1, T_A = 1, K_A = 190, V_{ref} = 1.05, E_{fd\,max} = 5, E_{fd\,min} = 0$ and when $D = 2.0$, the phase plane behaviours are shown in Figure 3, reflecting the chaotic behaviour of the system (6).

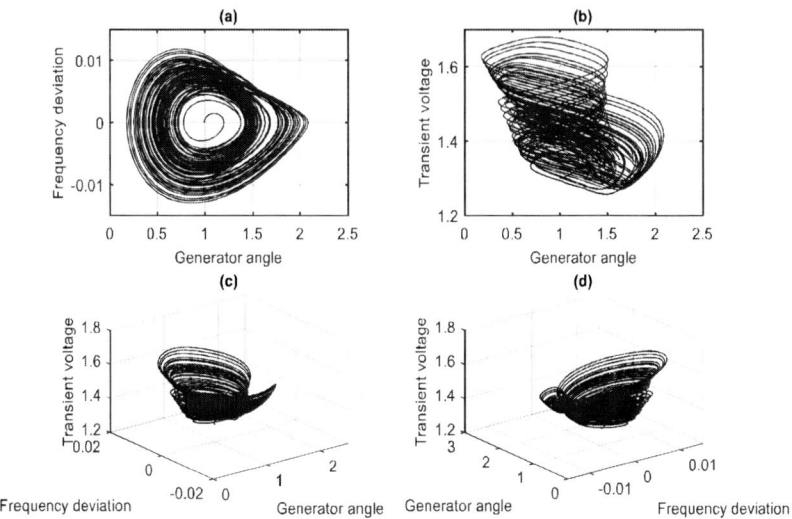

Figure 3. Phase plane behaviour of FDFPS shows chaotic nature.

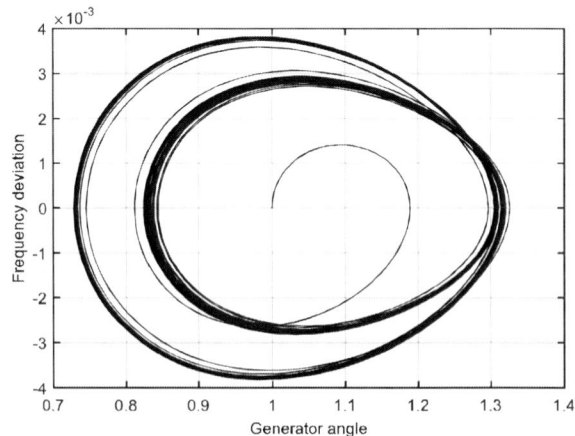

Figure 4. Phase plane behaviour shows the period-doubling bifurcation.

The Lyapunov exponents of the extended SMIB power system (6) are calculated as 0.1227, 0.0073 ≈ 0, −0.2374 and −1.374. One positive Lyapunov exponent confirms the chaotic nature of the extended SMIB power system. At $D = 5.1$, the phase plane behaviour is shown in Figure 4 demonstrates the period-doubling bifurcation nature of the extended SMIB power system.

In the next section, conventional, proportional-integral SMC techniques are summarised, and the design of the HOSMC technique is presented.

3. CONVENTIONAL, PI AND HIGHER-ORDER SMC DESIGN FOR 4D FUNDAMENTAL POWER SYSTEM

Sliding mode control design provides a systematic approach to maintain stability and consistent performance and eliminates the necessity of exact modelling [28]. The procedure of SMC involves two steps. In the first step, a sliding surface is designed such that while on the sliding surface, the plant dynamics is restricted onto the sliding surface, i.e., $s = 0$. In the second step, an appropriate control law is designed to guarantee the convergence of the system trajectory onto the sliding surface in a finite time, i.e., $\dot{s} = 0$.

Let, $\delta = x_1, \omega = x_2, E = x_3$ and $E_{fdr} = x_4$ be the states variables. After the mathematical calculation, the extended SMIB power system (6) in terms of the new variables $x \in \mathbb{R}^4$, is written as:

$$\begin{cases} \dot{x}_1 = c_1 x_2 \\ \dot{x}_2 = -c_2 x_2 + c_3 - c_4 x_3 \sin x_1 \\ \dot{x}_3 = -c_5 x_3 + c_6 \cos x_1 + c_7 \\ \dot{x}_4 = -c_8 \left(\sqrt{c_9 + c_{10} x_3 \cos x_1 + c_{11} x_3^2} \right) + c_{12} - c_{13} x_4 \end{cases} \quad (7)$$

where $c_1, c_2, \ldots, c_{13}$ are

$$\begin{cases} c_1 = 2\pi f_0 \\ c_2 = D/M \\ c_3 = P_m/M \\ c_4 = \frac{1}{(x'_d+x)M} \end{cases}, \begin{cases} c_5 = \frac{(x_d+x)}{(x'_d+x)T'_{d_0}} \\ c_6 = \frac{(x_d - x'_d)}{(x'_d+x)T'_{d_0}} \\ c_7 = E_{fdr}/T'_{d_0} \end{cases}, \begin{cases} c_8 = K_A/(x'_d + x)T_A \\ c_9 = x'^2_d \\ c_{10} = 2x'_d x \end{cases},$$

The tracking error is defined as:

$$e_i = x_i - r_i \text{ for } i = 1, 2, 3, 4 \qquad (8)$$

where r_1, r_2, r_3 and r_4 are the reference command.
The error dynamics is written as:

$$\dot{e}_i = \dot{x}_i - \dot{r}_i \qquad (9)$$

Our objective is to design an appropriate SMC controller u_i for $i = 1, 2, 3, 4$ such that tracking error dynamics (9) is asymptotically stable, i.e., $\lim_{t \to \infty} \|e(t)\| = 0 \ \forall \ e(t) \in R^4$. Following subsections include the conventional SMC, proposed pole placement based PISMC and HOSMC designs.

3.1. Conventional Sliding Mode Control Design for Extended SMIB Power System

Let switching surfaces be defined as follows:

$$s_i = e_i \qquad (10)$$

The dynamics of sliding surfaces are obtained as:

$$\dot{s}_i = \dot{e}_i \qquad (11)$$

Let the equivalent control laws be defined as:

$$\begin{cases} u_{eq1} = \dot{r}_1 - c_1 x_2 \\ u_{eq2} = \dot{r}_2 + c_2 x_2 - c_3 \\ u_{eq3} = \dot{r}_3 + c_5 x_3 - c_7 \\ u_{eq4} = \dot{r}_4 + c_8\left(\sqrt{c_9 + c_{10} x_3 \cos x_1 + c_{11} x_3^2}\right) - c_{12} + c_{13} x_4 \end{cases} \quad (12)$$

The system dynamics operate in sliding mode when $\dot{s}_i = 0$ for $i = 1, 2, 3, 4$. The reaching control laws are defined as:

$$\begin{cases} u_{r1} = -\zeta sign(s_1) \\ u_{r2} = -\zeta sign(s_2) \\ u_{r3} = -\zeta sign(s_3) \\ u_{r4} = -\zeta sign(s_4) \end{cases} \quad (13)$$

Therefore, the net control laws are written as $u_i = u_{eq_i} + u_{r_i}$ for $i = 1, 2, 3, 4$. Then we have

$$\begin{cases} u_1 = \dot{r}_1 - c_1 x_2 - \zeta sign(s_1) \\ u_2 = \dot{r}_2 + c_2 x_2 - c_3 - \zeta sign(s_2) \\ u_3 = \dot{r}_3 + c_5 x_3 - c_7 - \zeta sign(s_3) \\ u_4 = \dot{r}_4 + c_8\left(\sqrt{c_9 + c_{10} x_3 \cos x_1 + c_{11} x_3^2}\right) - c_{12} + c_{13} x_4 - \zeta sign(s_4) \end{cases}$$

$$(14)$$

The results using conventional SMC is presented in the result and discussion section. In the following subsection, pole placement based PISMC is discussed.

3.2. Proportional Integral Sliding Mode Control Design for Extended SMIB Power System

In the proposed control strategy, a pole placement based PISMC is designed. The proposed control strategy is presented in Figure 3.

The Jacobian linearisation gives the A and B matrices as:

$$A = \begin{bmatrix} 0 & c_1 & 0 & 0 \\ 0 & -c_2 & 0 & 0 \\ 0 & 0 & -c_5 & 0 \\ 0 & 0 & 0 & -c_{13} \end{bmatrix} \quad B = \begin{bmatrix} 1 & 0 & 0 & 0 \\ 0 & 1 & 0 & 0 \\ 0 & 0 & 1 & 0 \\ 0 & 0 & 0 & 1 \end{bmatrix}$$

The gain matrix L is chosen as $L = diag(1,1,1,1)$ such that $LB = diag(1,1,1,1)$ is non-singular. As per desired reference command, let desired closed-loop poles be $p = [-1, -1.5, -0.5, -2]$ result in an acceptable transient response.

The original closed system matrix $(A - BK)$, where K is state feedback control gain. Therefore, using the pole placement method, the gain matrix K is calculated as

$$K = \begin{bmatrix} 1 & 376.8 & 0 & 0 \\ 0 & 1 & 0 & 0 \\ 0 & 0 & 0.333 & 0 \\ 0 & 0 & 0 & 1 \end{bmatrix}.$$

As a result, the matrix $L(A - BK)$ is computed in (15).

$$L(A - BK) = diag(k_1, k_2, k_3, k_4) = diag(-1, -1.5, -0.5, -2) \quad (15)$$

Let the PI sliding surfaces based on the pole placement of poles be defined as $s_i = L(e_i + \int_0^t (A - BK) e(\tau) d(\tau))$.

Then, the sliding surface is defined as: =

$$\begin{cases} s_1 = e_1 + \int_0^t k_1 e_1(\tau) \, d\tau \\ s_2 = e_2 + \int_0^t k_2 e_2(\tau) \, d\tau \\ s_3 = e_3 + \int_0^t k_3 e_3(\tau) \, d\tau \\ s_4 = e_4 + \int_0^t k_4 e_4(\tau) \, d\tau \end{cases} \quad (16)$$

where $k_1 = -1, k_2 = -1.5, k_3 = -0.5$ and $k_4 = -2$ are the integral gains for different sliding surfaces.

Further, the sliding dynamics is written as:

$$\begin{cases} \dot{s}_1 = \dot{e}_1 + e_1 \\ \dot{s}_2 = \dot{e}_2 + 1.5e_2 \\ \dot{s}_3 = \dot{e}_3 + 0.5e_3 \\ \dot{s}_4 = \dot{e}_4 + 2e_4 \end{cases} \quad (17)$$

The system operates in the sliding mode when $\dot{s}_i = 0$ for $i \in \mathbb{R}^4$. To ensure the occurrence of sliding motion, the reaching control laws be defined as:

$$\begin{cases} u_{r1} = -\zeta_1 sign(s_1) \\ u_{r2} = -\zeta_2 sign(s_2) \\ u_{r3} = -\zeta_3 sign(s_3) \\ u_{r4} = -\zeta_4 sign(s_4) \end{cases} \quad (18)$$

The equivalent control law is designed as:

$$\begin{cases} u_{eq1} = \dot{r}_1 - c_1 x_2 - x_1 + r_1 \\ u_{eq2} = \dot{r}_2 + c_2 x_2 - c_3 - 1.5 x_2 + 1.5 r_2 \\ u_{eq3} = \dot{r}_3 + c_5 x_3 - c_7 - 0.5 x_3 + 0.5 r_3 \\ u_{eq4} = \dot{r}_4 + c_8 \left(\sqrt{c_9 + c_{10} x_3 \cos x_1 + c_{11} x_3^2} \right) - c_{12} + c_{13} x_4 - 2 x_4 + 2 r_4 \end{cases}$$

(19)

To ensure the occurrence of sliding motion, the net control laws $u_i = u_{eq_i} + u_{r_i}$ for $i \in \mathbb{R}^4$ are defined in (20).

$$\begin{cases} u_1 = \dot{r}_1 - c_1 x_2 - x_1 + r_1 - \zeta_1 sign(s_1) \\ u_2 = \dot{r}_2 + c_2 x_2 - c_3 - 1.5 x_2 + 1.5 r_2 - \zeta_2 sign(s_2) \\ u_3 = \dot{r}_3 + c_5 x_3 - c_7 - 0.5 x_3 + 0.5 r_3 - \zeta_3 sign(s_3) \\ u_4 = \dot{r}_4 + c_8 \left(\sqrt{c_9 + c_{10} x_3 \cos x_1 + c_{11} x_3^2} \right) \\ \quad\quad - c_{12} + c_{13} x_4 - 2 x_4 + 2 r_4 - \zeta_4 sign(s_4) \end{cases} \quad (20)$$

Therefore, the equivalent error dynamics is written as:

$$\begin{cases} \dot{e}_1 = -e_1 \\ \dot{e}_2 = -1.5e_2 \\ \dot{e}_3 = -0.5e_3 \\ \dot{e}_4 = -2e_4 \end{cases} \quad (21)$$

3.3. Design of Higher Order Sliding Mode Control for Extended SMIB Power System

The block diagram of the proposed HOSMC technique is shown in Figure 5.

$$\begin{cases} u_1 = 2\dot{r}_1 - 2c_1x_2 - x_1 + r_1 + -\ddot{x}_1 + \ddot{r}_1 - \zeta_{11}sign(s_1) - \zeta_{21}sign(\dot{s}_1) \\ u_2 = 2\dot{r}_2 + 2c_2x_2 - 2c_3 - 1.5x_2 + 1.5r_2 \pm 0.67\ddot{x}_2 + 0.67\ddot{r}_2 \\ \qquad -\zeta_{22}sign(\dot{s}_2) - \zeta_{12}sign(s_2) \\ u_3 = 2\dot{r}_3 + 2c_5x_3 - 2c_7 - 0.5x_3 + 0.5r_3 + \ddot{x}_3 - \ddot{r}_3 - \zeta_{13}sign(s_3) \\ \qquad -\zeta_{23}sign(\dot{s}_3) \\ u_4 = 2\dot{r}_4 + 2c_8\left(\sqrt{c_9 + c_{10}x_3 \cos x_1 + c_{11}x_3^2}\right) - 2c_{12} + 2c_{13}x_4 \\ \qquad -2x_4 + 2r_4 - 0.5\ddot{x}_4 + 0.5\ddot{r}_4 - \zeta_{14}sign(s_4) - \zeta_{24}sign(\dot{s}_4) \end{cases}$$

$$(22)$$

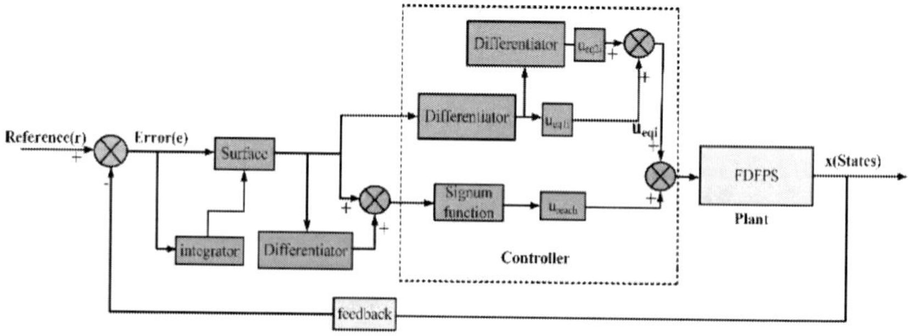

Figure 5. Block diagram of pole placement based HOSMC control strategy.

Theorem 1. The designed control law in (22) ensures the reachability condition or existence of a stable sliding surface, i.e., $[s\dot{s} + \dot{s}\ddot{s} \leq 0]$.

Proof. Let a Lyapunov function candidate $V(s)$ be defined as:

$$V(s_i) = \frac{1}{2}s_i^2 + \frac{1}{2}\dot{s}_i^2 \text{ for } i = 1, 2, 3, 4 \tag{23}$$

Assuming $V(s)$ is a continuously differentiable function to ensure a stable sliding surface.

Then, for the sliding surface s_1, $V(s_1) = 0.5s^2{}_1 + 0.5\dot{s}^2{}_1$

$$\dot{V}(s_1) = s_1\dot{s}_1 + \dot{s}_1\ddot{s}_1$$

$$\dot{V}(s_1) = s_1[\dot{e}_1 + e_1] + \dot{s}_1[\ddot{e}_1 + \dot{e}_1]$$

$$\dot{V}(s_1) = s_1[c_1 x_2 + u_{11} - \dot{r}_1 + x_1 - r_1] + \dot{s}_1[\ddot{x}_1 - \ddot{r}_1 + c_1 x_2 + u_{21} - \dot{r}_1]$$

Using the value of u_{11} and u_{21} we have:

$$\dot{V}(s_1) = s_1[c_1 x_2 + \dot{r}_1 - c_1 x_2 - x_1 + r_1 - \zeta_{11}sign(s_1) - \dot{r}_1 + x_1 - r_1]$$
$$+ \dot{s}_1[\ddot{x}_1 - \ddot{r}_1 + c_1 x_2 + \dot{r}_1 - \ddot{x}_1 + \ddot{r}_1 - c_1 x_2 - \dot{r}_1$$
$$- \zeta_{21}sign(\dot{s}_1)]$$

$$\dot{V}(s_1) = -[s_1\zeta_{11}sign(s_1)] - [\dot{s}_1\zeta_{21}sign(\dot{s}_1)]$$

Reachability condition exists when $|\zeta_{11}| > 0$ and $|\zeta_{21}| > 0$.

For sliding surface s_2:

$$V(s_2) = 0.5s^2{}_2 + 0.5\dot{s}^2{}_2$$

$$\dot{V}(s_2) = s_2\dot{s}_2 + \dot{s}_2\ddot{s}_2$$

$$\dot{V}(s_2) = s_2(\dot{e}_2 + 1.5e_2) + \dot{s}_2(\ddot{e}_2 + 1.5\dot{e}_2)$$

$$\dot{V}(s_2) = s_2(-c_2x_2 + c_3 - c_4x_3 \sin x_1 + u_{12} - \dot{r}_2 + 1.5x_2 - 1.5r)$$
$$+ \dot{s}_2[\ddot{x}_2 - \ddot{r}_2 + 1.5(-c_2x_2 + c_3 - c_4x_3 \sin x_1 - \dot{r}_2$$
$$+ u_{22}) - 1.5\ddot{r}_2]$$

By substituting the values of u_{12} and u_{22}, we get $\dot{V}(s_2)$ as:

$$\dot{V}(s_2) = s_2(-c_2x_2 + c_3 - c_4x_3 \sin x_1 + c_2x_2 - c_3 + \dot{r}_2 - 1.5x_2 + 1.5r_2 - \zeta sign(s_2) - \dot{r}_2 + 1.5x_2 - 1.5r_2) + \dot{s}_2(\ddot{x}_2 - \ddot{r}_2 + 1.5(-c_2x_2 + c_3 - c_4x_3 \sin(x_1) - \dot{r}_2 + -0.67\ddot{x}_2 + 0.67\ddot{r}_2 + c_2x_2 - c_3 + \dot{r}_2) - 1.5\ddot{r}_2)$$

$$\dot{V}(s_2) = s_2[-c_4x_3 \sin x_1 - \zeta_{21}sign(s_2)] + \dot{s}_2[-c_4x_3 \sin x_1 - \zeta_{22}sign(\dot{s}_2)]$$

Now reachability condition exists when $|\zeta_{12}| > |c_4x_3 \sin x_1|$ and $|\zeta_{22}| > |c_4x_3 \sin x_1|$.

Similarly, for surface s_3:

$$\dot{V}(s_3) = s_3\dot{s}_3 + \dot{s}_3\ddot{s}_3$$

$$\dot{V}(s_3) = s_3(\dot{e}_3 + 0.5e_3) + \dot{s}_3(\ddot{e}_3 + 0.5\dot{e}_3)$$

$$\dot{V}(s_3) = s_3(-c_5x_3 + c_6 \cos x_1 + c_7 + u_{13} - \dot{r}_3 + 0.5x_3 - 0.5r_3) + \dot{s}_3(\ddot{x}_3 - \ddot{r}_3 + 0.5(-c_5x_3 + c_6 \cos x_1 + c_7 + u_{23}) - 0.5\ddot{r}_3)$$

Using the values of u_{13} and u_{23}, $\dot{V}(s_3)$ is written as:

$$\dot{V}(s_3) = s_3(-c_5x_3 + c_6 \cos x_1 + c_7 + c_5x_3 - c_7 + \dot{r}_3 - 0.5x_3 + 0.5r_3 - \zeta_{13} sign(s_3) - \dot{r}_3 + 0.5x_3 - 0.5r_3) + \dot{s}_3(\ddot{x}_3 - \ddot{r}_3 + 0.5(-c_5x_3 + c_6 \cos x_1 + c_7 - 2\ddot{x}_3 + 2\ddot{r}_3 + c_5x_3 - c_7 + \dot{r}_3) - 0.5\ddot{r}_3 - \zeta_{23} sign(\dot{s}_3))$$

$$\dot{V}(s_3) = s_3[c_6 \cos x_1 - \zeta_{13} sign(s_3)] + \dot{s}_3[c_6 \cos x_1 - \zeta_{23} sign(\dot{s}_3)]$$

Now reachability condition exists when $|\zeta_{13}| > |c_6 \cos x_1|$ and $|\zeta_{23}| > |c_6 \cos x_1|$.

Similarly, for surface s_4:

$$\dot{V}(s_4) = s_4 \dot{s}_4 + \dot{s}_4 \ddot{s}_4$$

$$\dot{V}(s_4) = s_4(\dot{e}_4 + 2e_4) + \dot{s}_4(\ddot{e}_4 + 2\dot{e}_4)$$

$$\dot{V}(s_4) = s_4(-c_8\left(\sqrt{c_9 + c_{10}x_3 \cos x_1 + c_{11}x_3^2}\right) + c_{12} - c_{13}x_4 + u_{14} - \dot{r}_4 + 2x_4 - 2r_4) + \dot{s}_4(\ddot{x}_4 - \ddot{r}_4 + 2(-c_8\left(\sqrt{c_9 + c_{10}x_3 \cos x_1 + c_{11}x_3^2}\right) + c_{12} - c_{13}x_4 + u_{24}) - 2\dot{r}_4)$$

By substituting the value of u_{14} and u_{24}, results $\dot{V}(s_4)$ as:

$$\dot{V}(s_4) = s_4[-\zeta_{14}\, sign(s_4)] + \dot{s}_4[-\zeta_{24}\, sign(\dot{s}_4)]$$

Reachability condition exists when $|\zeta_{14}| > 0$ and $|\zeta_{24}| > 0$.

Therefore, Theorem 1 ensures the stability of designed sliding surfaces and convergence of state trajectories onto the sliding surfaces.

4. COMPARATIVE PERFORMANCE ANALYSIS AND DISCUSSION

Simulation is conducted in MATLAB Simulink and runs for 100 seconds with variable steps. The design variable for the sliding surface is $\zeta_{11} = \zeta_{12} = 0.1$, $\zeta_{21} = \zeta_{22} = 0.2$, $\zeta_{13} = \zeta_{23} = 0.1$ and $\zeta_{14} = \zeta_{24} = 0.1$. The initial condition $[x_1(0)\ x_2(0)\ x_3(0)\ x_4(0)]^T$ for the state variables of the proposed power system is considered as $[1.0\ 0\ 0\ 1.3\ 1.9]^T$.

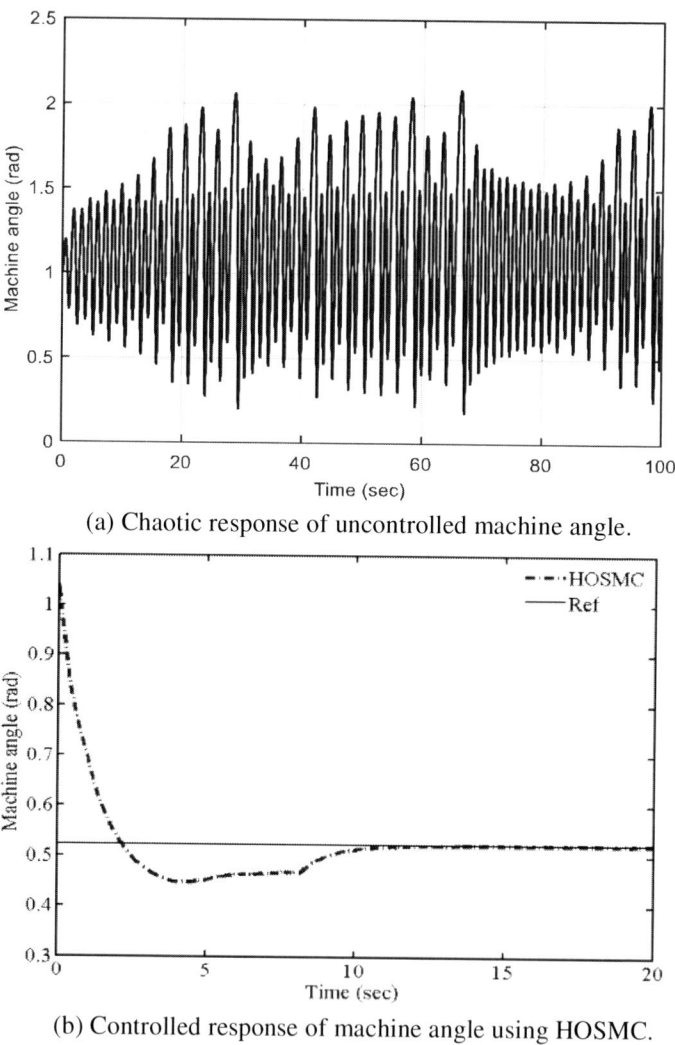

(a) Chaotic response of uncontrolled machine angle.

(b) Controlled response of machine angle using HOSMC.

Figure 6. Chaotic and controlled response of machine angle using HOSMC.

4.1. Simulation Results for HOSMC Design

The chaotic nature of the machine angle (δ) is shown in Figure 6(a). The controlled response of machine angle (δ) is shown in Figure 6(b). Figure 6(b) shows the HOSMC based controlled response of machine angle, which shows smooth control with a settling time of 11 seconds approximately.

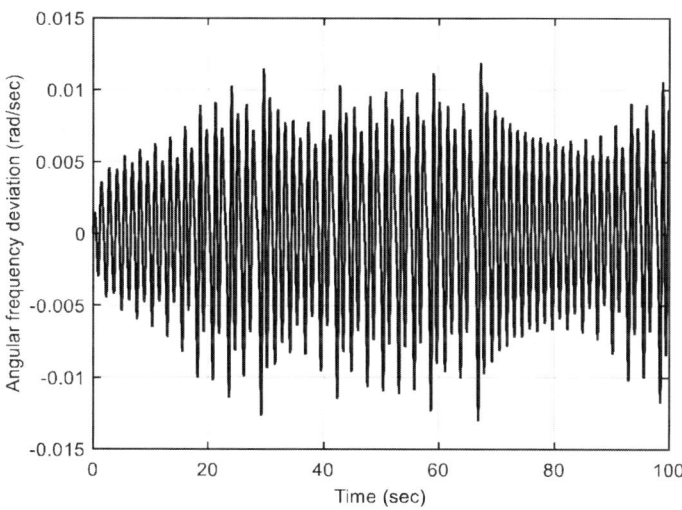

(a) Chaotic response of angular frequency deviation.

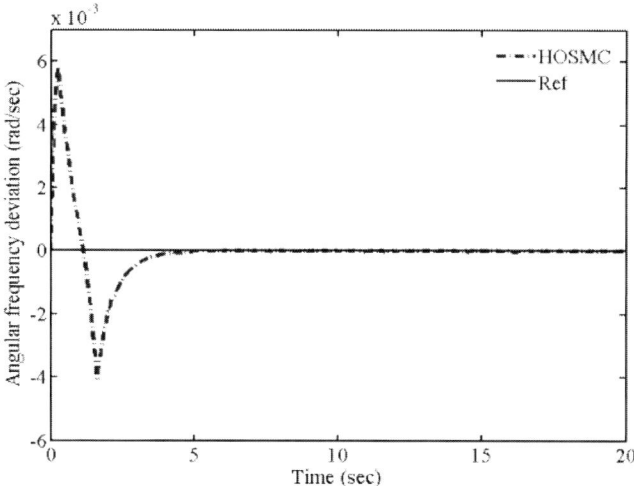

(b) HOSMC based controlled angular frequency deviation.

Figure 7. Chaotic and HOSMC based controlled responses of angular frequency deviation.

The angular frequency (ω) is shown in Figure 7 for chaotic and controlled responses. Figure 7(b) reveals that angular frequency is controlled to the desired or reference value, and angular frequency deviation is controlled with the accuracy up to the order of 10^{-3}.

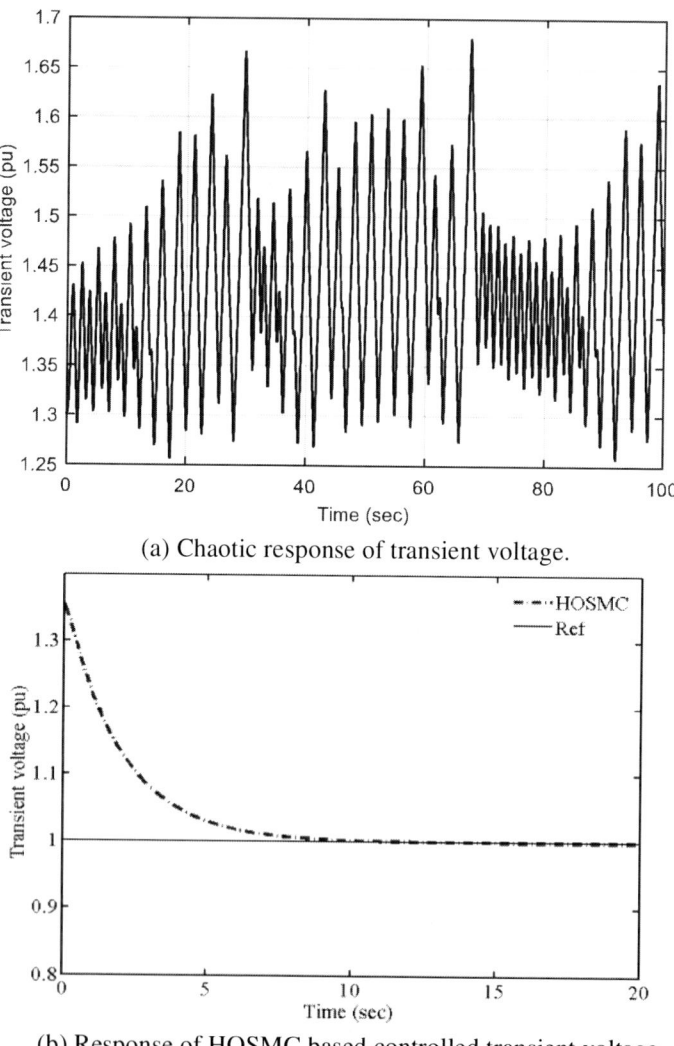

(a) Chaotic response of transient voltage.

(b) Response of HOSMC based controlled transient voltage.

Figure 8. Chaotic and higher-order sliding mode-controlled responses of transient voltage.

Chaos Control in Power System ... 103

The actual or chaotic and controlled behaviours of transient voltage (E') are shown in Figure 8. Figure 8(b) shows the controlled response of transient voltage using HOSMC and confirms that the chaotic oscillation is controlled and settles to the reference point in approximately 10 seconds.

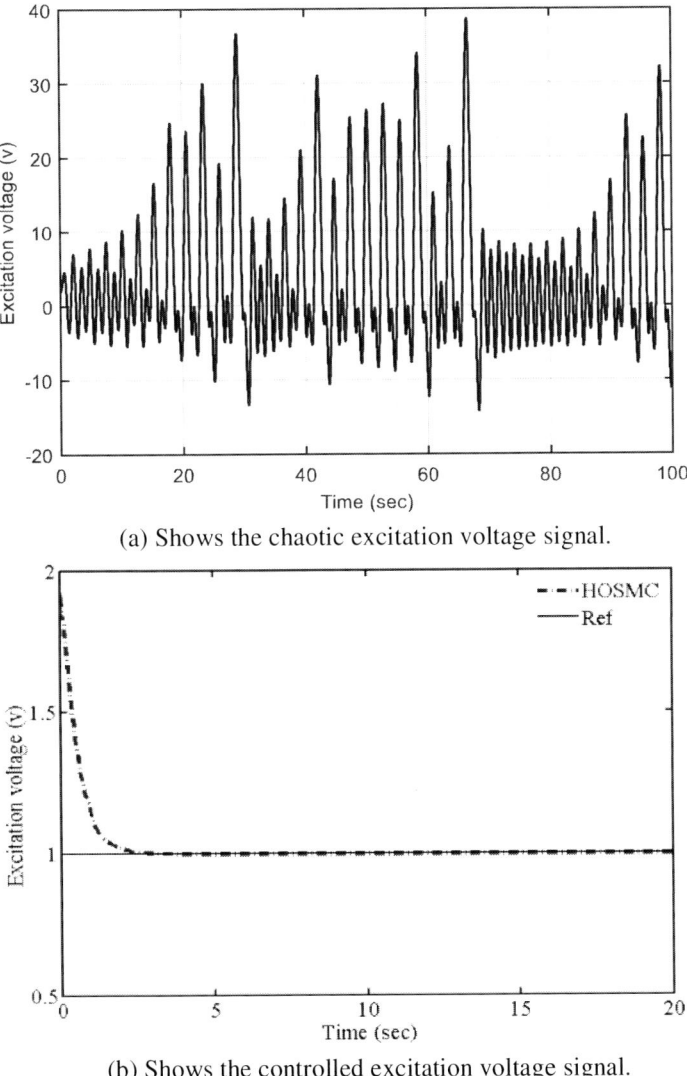

(a) Shows the chaotic excitation voltage signal.

(b) Shows the controlled excitation voltage signal.

Figure 9. The chaotic and HOSMC based controlled excitation voltage signals.

Similarly, the chaotic and controlled response of excitation voltage (E_{fdr}) are shown in Figure 9. The uncontrolled or chaotic response of excitation voltage is depicted in Figure 9(a), which shows that the nature of the excitation voltage is chaotic. Figure 9(b) shows that the chaotic response of excitation voltage is controlled successfully using the HOSMC technique. The transient response of controlled excitation voltage shows smooth characteristics.

4.2. Comparison among Three SMC Control Techniques

Controlled responses of machine angle (δ) are shown in Figure 10, which reflects the machine angle is controlled using conventional SMC, PISMC and HOSMC towards the desired reference value. It is apparent that the reference value is obtained in approximately 10 second. Figure 10 reflects that the PISMC based controlled response is faster (approximately 6 seconds) than the classical SMC (about 8 seconds), with a comparatively smooth appearance. HOSMC based controlled response of machine angle is comparatively smooth appearance than classical SMC and PISMC techniques at the price of higher settling time (approximately 10 seconds).

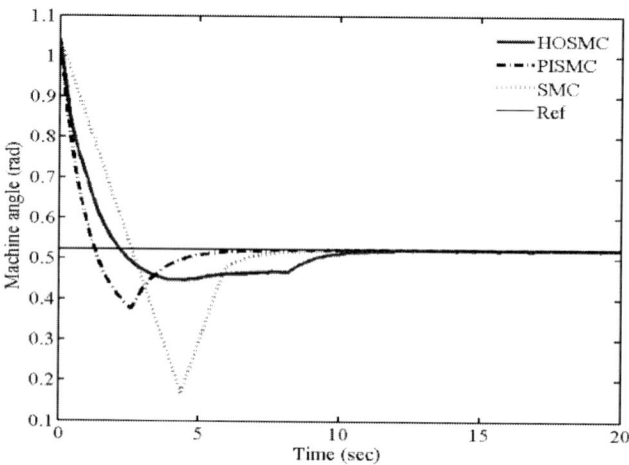

Figure 10. The response of controlled machine angle using classical SMC, PISMC and HOSMC techniques.

Figure 11 reveals that angular frequency deviation is controlled to the desired reference value. It is apparent that the desired value is obtained successfully. The HOSMC based controlled response is better than the classical SMC and PISMC with the least overshoot and settling time (approximately 2 seconds).

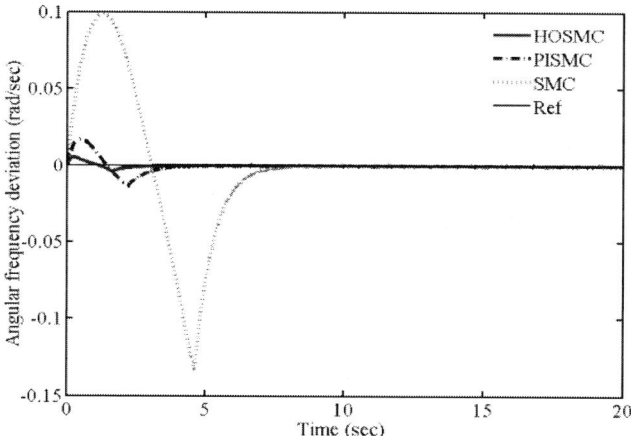

Figure 11. The controlled response of the angular frequency deviation using classical SMC, PISMC and HOSMC techniques.

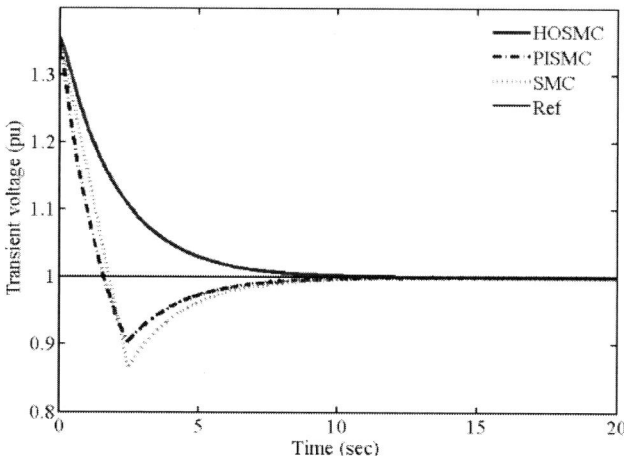

Figure 12. The controlled transient voltage using three SMC techniques.

The transient voltage is controlled as per the desired reference value and shown in Figure 12. It is apparent that the desired reference value is reached in approximately 10 seconds. Oscillations are not present using the HOSMC technique and working well than the other two SMC techniques.

The controlled excitation voltage using conventional SMC, PISMC and HOSMC is depicted in Figure 13. It is apparent that the excitation voltage is controlled to the reference value in approximately 10-12 seconds using the three control techniques. Figure 13 reveals that the HOSMC works better with the smooth feature of excitation voltage.

It is observed that commenting on the suitable control technique is not possible as per the results shown in Figure 6 to Figure 9. Therefore, two measures are considered. Two measures are average error energy and average control energy.

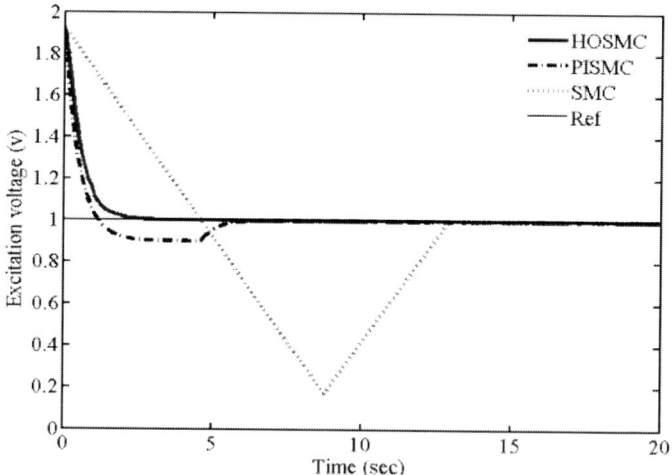

Figure 13. Shows the controlled excitation voltage using classical SMC, PISMC and HOSMC techniques.

Table 1. Comparison among Different SMC Techniques

Controller	E_e	E_u
Conventional Sliding Mode Control (SMC)	6.260	2.670e4
Proportional Integral Sliding Mode Control (PISMC)	1.085	2.240e4
Higher Order Sliding Mode Control (HOSMC)	1.400	2.076e4

4.3. Quantitative Analysis

The average error (E_e) and average control (E_u) energies are defined as $E_e = \int_0^t \sqrt{e_1^2 + e_2^2 + e_3^2 + e_4^2}\, dt$ and $E_u = \int_0^t \sqrt{u_1^2 + u_2^2 + u_3^2 + u_4^2}\, dt$ and calculated for the three SMC controllers under a similar scenario (refer to Table 1). It is observed that average control energy corresponding to the HOSMC technique is less average control energy as compared to the SMC and PISMC techniques, but there is a trade-off between average error and average control energies for PISMC and HOSMC techniques. As a result, the comparison among SMC techniques cannot be concluded. Therefore, calculation of the performance indices of the SMC techniques is the possible way for comparison.

Table 2. Performance Indices Analysis among Different SMC Techniques

Techniques (M)	Performance indices (J_M^C)		
	Case 1	*Case 2*	*Case 3*
SMC	2.13e4	1.33e4	0.53e4
PISMC	1.79e4	1.12e4	0.44e4
HOSMC	1.66e4	1.03e4	0.41e4

Three feasible performance indices based on different weights of average error and average control energies are considered and defined as:

Case 1: Weightage of E_e 20% & E_u 80%.

$$J_M^1 = \left(\frac{1}{5}E_e + \frac{4}{5}E_u\right)$$

Case 2: Weightage of E_e 50% & E_u 50%.

$$J_M^2 = \left(\frac{1}{2}E_e + \frac{1}{2}E_u\right)$$

Case 3: Weightage of E_e 80% & E_u 20%.

$$J_M^3 = \left(\frac{4}{5}E_e + \frac{1}{5}E_u\right)$$

where J_M^C is the performance index. Superscript '$C = 1, 2, 3$' stand for three *Cases* and subscript 'M' represents three SMC methods or techniques. The above three cases are analysed and presented in Table 2. The calculated performance indices reveal that the performance of the HOSMC technique is better than the other two SMC techniques.

CONCLUSION AND FUTURE SCOPE

In this chapter, chaos in the extended SMIB power system is reported, and chaos control is achieved. A higher-order sliding mode control technique is designed to suppress chaos employing pole placement. In the proposed HOSMC technique, proportional integral switching surfaces are devised using the appropriate placement of closed-loop poles. The designed HOSMC laws ensure the safe and stable operation of the extended SMIB power system and reduce the chattering in the system states. Finally, MATLAB results substantiate the proposed pole placement based HOSMC technique. Simulation results reveal that the design of the proposed controller reduces chaos in the extended SMIB power system. Based on three performance indices, a comparative performance analysis of conventional SMC, PISMC and HOSMC techniques is achieved. The performance of HOSMC is better than the other two SMC techniques. HOSMC technique is superior in the implementation of control signal because the control effort by the HOSMC technique is minimum among the three sliding mode control techniques.

REFERENCES

[1] Wildberger M., Stability and nonlinear dynamics in power systems, *IEEE Power Engineering Review*, 14, 16-18, (1994).

[2] Kundur P. and Taylor C., Blackout Experiences and Lessons, Best Practices for System Dynamic Performance, the Role of new Technologies, *IEEE Task Force Report*, 2007.

[3] Nayfeh H., Harb A. M. and Chin C., Bifurcations in a power system model, *IEEE International Symposium on Circuits and Systems*, Seattle, 1995, 283-286.

[4] Ji W. and Venkatasubramaniam V., Hard-limit induced chaos in a power system model, *Int. J. Electrical Power & Energy Systems*, 18, 279-295, (1996).

[5] Harb M. and Jabbar N. A., Controlling Hopf bifurcation and chaos in a small power system, *Journal of Power and Energy Engineering*, 18, 1055-1063, (2003).

[6] Harb M. and Widyan M. S., Chaos and bifurcation control of SSR, *IEEE Second benchmark model*, 21, 537-552, (2004).

[7] McClamroch N. H., Nonlinear Control of Engineering Systems: A Lyapunov-Based Approach, *IEEE Control Systems*, 25:2, 79-88, (2005).

[8] Chen H. K., Lin T. N. and Chen J. H., Dynamic analysis, controlling chaos and chaotification of a SMIB power system, *Chaos, Solitons & Fractals*, 24, 1307-1315, (2005).

[9] Ren H. and Liu D., Nonlinear feedback control of chaos in permanent magnet synchronous motor, *IEEE Trans. Circuits and Systems*, 53, 45-50, (2006).

[10] Shahverdiev E. M., Hashimova L. H. and Hashimova N. T., Chaos synchronisation in some power systems, *Chaos, Solitons & Fractals*, 37, 827-834, (2008).

[11] Ginarsa M., Soeprijanto A. and Purnomo M. H., Controlling chaos using ANFIS-based Composite Controller (ANFIS-CC) in power systems, *Int. Conference on Instrumentation, Communication, Information Tech., and Biomedical Eng.*, Bandung, (2009), 1-5.

[12] Khaki, Noroozi N. and Seifi A., Chaotic Oscillations Damping in Power System by Adaptive Control Theory, *ICEE Iranian conference on electrical engineering*, 7, 351-356, (2009).

[13] Wei Q. and Qin Y. H., Controlling Chaos in Single-Machine-Infinite Bus Power System by Adaptive Passive Method, *Fourth International Workshop on Chaos-Fractals Theories and Applications*, Hangzhou, (2011), 295-297.

[14] Ghaffari V. and Shabaninia F., Application of extended Kalman filter to chaos synchronisation, *Third Int. Conference on Control, Instrumentation, and Automation*, Tehran, (2013), 71-76.

[15] Ma M., Min F., Huang W. and Cao Y., Dynamic analysis and controlling chaos of the electrical power system with the disturbance of electromagnetic power, *Proc. 33rd Chinese Control Conference*, Nanjing, (2014), 1994-1998.

[16] Cleju M., About robust control on nonlinear chaotic oscillators, *International Conference and Exposition on Electrical and Power Engineering (EPE)*, Iasi, (2014), 042-046.

[17] Li, Liu L., Liu C., Hu X. and Li A., Chaos suppression for a four-dimensional fundamental power system model using adaptive feedback control, *Trans. Institute Measurement and Control*, 1-14, (2015).

[18] Singh K. and Pal B. C., Decentralized Control of Oscillatory Dynamics in Power Systems Using an Extended LQR, *IEEE Trans. Power Syst.*, 31:3, 1715-1728, (2016).

[19] Do T. D., Optimal control design for chaos suppression of PM synchronous motors, *2nd International Conference on Control Science and Systems Engineering (ICCSSE)*, Singapore, (2016), 88-92.

[20] Liang Z. and Gao J., Chaos in a Fractional-Order Single-Machine Infinite-Bus Power System and Its Adaptive Backstepping Control, *Int. J. Modern Nonlinear Theory and Application,* 5, 122-131, (2016).

[21] Singh K. and Pal B. C., An extended linear quadratic regulator for LTI systems with exogenous inputs, *Automatica*, 76, 10-16, (2017).

[22] Ni, Liu L., Liu C., Hu X. and Li S., Fast Fixed-Time Nonsingular Terminal Sliding Mode Control and Its Application to Chaos Suppression in Power System, *IEEE Trans. Circuits and Systems-I: Regular Papers*, 64:2, 151-155, (2017).

[23] Liu H. Ye. Y., Zhang P. and Du Z., Analysis and Detection of Forced Oscillation in Power System, *IEEE Trans. Power Syst.*, 32:2, 1149-1160, (2017).

[24] Das S. and Yadav V. K., Stability analysis, chaos control of fractional order Vallis and El-Nino systems and their synchronisation, *IEEE/CAA J. Autom. Sinica*, 4:1, 114-124, (2017).

[25] Singh K. and Pal B. C., Decentralized Nonlinear Control for Power Systems Using Normal Forms and Detailed Models, *IEEE Trans. Power Syst*, 33:2, 1160-1172, (2018).

[26] Mahmoud S. and Oyedeji M. O., Adaptive and predictive control strategies for wind turbine systems: a survey, *IEEE/CAA J. Autom. Sinica*, 6:2, 364-378, (2019).

[27] Wang Q. et al., Theoretical Design and FPGA-Based Implementation of Higher-Dimensional Digital Chaotic Systems, *IEEE Transaction on Circuits and Systems-I: Regular Papers*, 63:3, 401-412, (2016).

[28] Utkin V. I., Sliding mode control design principles and applications to electric drives, *IEEE Trans. Industrial Electronics*, 40, 23-36, (1993).

[29] Azlan Z., Osman J. H. S. and Ahmad M. N., Proportional Integral Sliding Mode Control of Hydraulic Robot Manipulators with Chattering Elimination, *First International Conference Industrial and Information Systems*, Peradeniya, (2006), 279-284.

[30] Ismail F., A reduce chattering problem using composite nonlinear feedback and proportional integral sliding mode control, *10th Asian Control Conference (ASCC)*, Kota Kinabalu, (2015), 1-6.

[31] Kang S. W. and Kim K. H., Sliding mode harmonic compensation strategy for power quality improvement of a grid-connected inverter under distorted grid condition, *IET Power Electronics*, 8, 1461-1472, (2015).

[32] Singh P. and Roy B. K., Comparative performances of synchronisation between different classes of chaotic systems using three control techniques, *Annual Reviews in Control*, 45, 152-165, (2018).

[33] Singh P. and Roy B. K., Comparative performances of anti-synchronisation between different chaotic systems using three control schemes, In: *Nonlinear Systems: Design, Applications and Analysis*, 1st ed., C. K. Volos, Ed. USA: Nova Science, (2017), 101-146.

[34] Guo J. and Dong L., Robust load frequency control for uncertain nonlinear interconnected power systems, *Int. J. Automation and Control*, 11:3, 239-261, (2017).

[35] Kumar M. and Singh P. P., Chaos control of a four-dimensional fundamental power system using pole placement based proportional integral sliding mode control, *Int. J. Automation and Control*, 13:6, 679-697, (2019).

[36] Morshed M. J. and Fekih A., A Sliding Mode Approach to Enhance the Power Quality of Wind Turbines Under Unbalanced Voltage Conditions, *IEEE/CAA J. Autom. Sinica*, 6:2, 566-574, (2019).

[37] Pal B. and Chaudhuri B., *Robust Control in Power Systems*. New York, USA: Springer, 2005, ch. 4.

In: Nonlinear Systems
Editor: P. P. Singh
ISBN: 978-1-68507-660-3
© 2022 Nova Science Publishers, Inc.

Chapter 5

CHAOS CONTROL IN THE ROTATIONAL MACHINE WITH CENTRIFUGAL FLYBALL GOVERNOR SYSTEM USING BACKSTEPPING SLIDING MODE CONTROL

Piyush Pratap Singh[*] *and Binoy Krishna Roy*[†]
Department of Electrical Engineering,
National Institute of Technology Meghalaya,
Shillong, Meghalaya, India
National Institute of Technology Silchar,
Cachar, Assam, India

Abstract

This chapter puts forward the control of chaos in the rotational machine with centrifugal flyball governor system. A new controller based on recursive backstepping sliding mode control (RBSMC) is proposed such that it can control the chaotic dynamics in the rotational machine with centrifugal flyball governor system to stabilize at any position or to track any trajectory that is a smooth function of time. A proportional integral switching surface is proposed to

---
[*]Corresponding Author's E-mail: piyushpratapsingh@gmail.com.
[†]Corresponding Author's E-mail: bkr_nits@yahoo.co.in.

achieve the stability condition of the error dynamics. Unlike the open loop and open plus closed loop control techniques, the design of proposed controller does not require the parameter perturbation. The required stability condition is derived based on Lyapunov stability theory. Simulation is achieved in MATLAB environment. Numerical simulation results are presented in order to show the effective verification of the proposed controller design. Simulation results correspond that the objective of chaos control is achieved successfully.

1. INTRODUCTION

During the past two decades, many studies have shown that chaotic phenomena are observed in many physical systems that possess non-linearity [1, 2, 3]. The centrifugal governor is a device that automatically controls the speed of an engine and prevents the damage caused by a sudden change of load torque. An autonomous hexagonal centrifugal governor system plays an important role in many rotational machines such as diesel engine, steam engine and hydro-turbine governing system [4, 5, 6]. When an engine system is subjected to external disturbances, the speed of the engine will vary. The dynamic system of the rotational machine with centrifugal governor exhibits a rich variety of nonlinear behaviours as certain parameters varied and because of nonlinearity, regular or chaotic motions may occur [7]. In order to decrease the change of engine speed and to avoid chaotic motion emerging in the operational process of the engine, the analysis, control and synchronization of such autonomous and non-autonomous rotational machines become the primary motivation of this research.

Various analytical and computational methods have been employed to study the dynamical behaviours of such nonlinear system. Anticontrol and synchronization of chaos for an autonomous rotational machine system with a hexagonal centrifugal governor has been discussed [8]. Linear and nonlinear controllers with certain feedback gain are proposed to anti-control the governor system is discussed. Linear feedback, nonlinear feedback, adaptive feedback, backstepping design and parameter evaluation from time sequences approaches are also used for synchronization of two coupled chaotic system. Control, anti-control and synchronization of chaos for an autonomous rotational machine system with time-delay is investigated in [9]. Phase diagram and power spectrum are presented to observe periodic and chaotic motions by applying numerical results. Linear feedback control and adaptive control algorithm are used to control chaos effectively. In the paper [10], the complex dynamic behaviours of the centrifugal flywheel governor systems are studied. The stability of the equilibrium points in the centrifugal flywheel governor system is explored and it is reported that the system has a rich variety of non-linear behaviours by numerically integrating the Lagrangian equations of motion. Global synchronization of non-autonomous rotational system with a centrifugal flywheel governor under linear state error feedback control is studied in [11].

In the recent decade, Hopf bifurcations, Lyapunov exponents and control of chaos for a class of centrifugal flywheel governor system is reported and hyperchaotic behaviour is observed in system [12] and Hopf bifurcation in a hexagonal governor system with a spring is investigated [13]. The dynamical equation and the state equation of mechanical centrifugal governor system are derived through Lagrange equations and the chaos anti-control method is proposed by addition of linear feedback [14]. The problem of robust finite time synchronization of non-autonomous chaotic centrifugal flywheel governors is discussed [15]. In this paper, synchronization is achieved under the effects of unknown parameters, model uncertainties and external disturbances. Finite time stabilization of non-autonomous uncertain chaotic centrifugal flywheel governor systems with input nonlinearities is reported [16]. Synchronization of the non-autonomous centrifugal flywheel governor systems using coupled feedback control and adaptive control methods is discussed [17]. Retrofication of mechanical speed governor with electronic speed governor for heavy duty diesel engines is reported [18]. Chaos suppression of the mechanical centrifugal flywheel governor system with output constraint and fully unknown parameters is investigated [19]. In this paper, a dynamic surface control scheme with neural network and tracking differentiator is developed by using tangent barrier Lyapunov function, to transform chaos oscillation into regular motion. Synchronization problem of chaotic centrifugal flywheel governor with parameters uncertainty and lumped disturbances is discussed [20]. A slave centrifugal flywheel governor system is considered as an under actuated system which a control input is designed to follow a master centrifugal flywheel governor system. The global dynamical behaviours of a hexagonal centrifugal governor with a spring are expanded to identify how the complex mode locking behaviours arise with the change in parameters [21]. By calculating the rotation number, it is unfolded that the interesting "Devils staircase" is presented with the parameter changed. Simultaneously, many control schemes has been reported for chaos control and synchronization in various field [22, 23, 24, 25, 26, 27, 28] etc.

Literature reveals that the purpose of backstepping design methodology is the construction of various types of control Lyapunov functions: stable, adaptive, robust, etc. The open loop and closed loop control techniques are employed to achieve the chaos control. The major limitations of the open loop and open plus closed loop control techniques are the adjustment of the parameters. Dynamic surface control was similar to backstepping and multiple surface control methods, but with an important addition, one low pass filter was included in the design which ended the complexity arising due to the "explosion of terms" that had made backstepping methods difficult to implement in practice. In view of several applications of the centrifugal flyball governor model and motivated with above literature review and issues, the present work is about the design of a recursive backstepping sliding mode control such that it can control and stabilize the chaotic dynamics of a rotational machine with centrifugal governor at any position or to track any trajectory that is a smooth function of time. Therefore, recursive backstepping sliding mode control technique (RBSMC) is investigated to mitigate the above issues. The proposed RBSMC technique is also more flexible than dynamic surface control. We could not find any such paper in the literature. Moreover, we want to emphasize your attention to the following points regarding the original contributions and novelty.

1. Backstepping technique usually adopts back step design to construct the Lyapunov function gradually, and then to design the corresponding virtue controller. The backstepping technique based on error also adopts back step design process, but the design of virtue controllers, depends on the corresponding errors, are designed to satisfy some expected behaviors. Therefore, sliding surface based errors are designed to the backstepping control design.

2. Various open loop and closed loop control techniques are reported to achieve the chaos control. The major limitations of the open loop and open plus closed loop control techniques is the adjustment of the

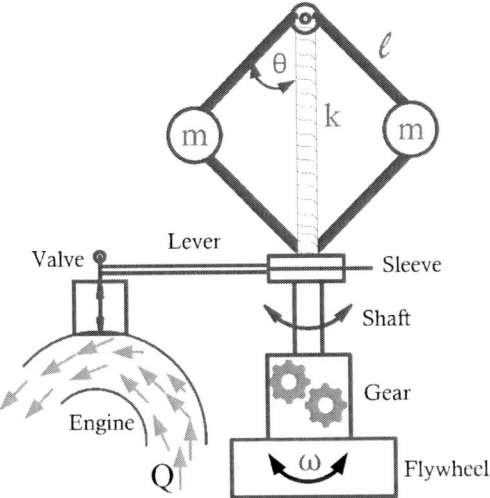

Figure 1. Physical model of centrifugal flyball governor system.

parameters. Therefore, recursive backstepping sliding mode control technique is investigated to mitigate the limitations.

3. Recursive backstepping sliding mode control (RBSMC) technique to control the rotational machine with centrifugal flyball governor is investigated first time in the literature.

Description of rotational machine with centrifugal flyball governor system model and the chaotic motion emerging in the operational process of the engine is discussed in detail in the next section.

2. DESCRIPTION OF ROTATIONAL MACHINE WITH CENTRIFUGAL FLYBALL GOVERNOR SYSTEM MODEL

The rotational machine with centrifugal flyball governor is depicted in Fig. 1. If the mass of the rods and sleeve are neglected, and the viscous damping in rod bearing of the flyball is represented by damping constant (c), then the equation of motion using the kinetic and potential energies of the governor system by Lagrange equation is derived in [7] and reproduced here as follows:

$$ml^2\ddot{\phi} - mrl\eta^2\,cos\phi - (2k + m\eta^2)l^2 sin\phi\,cos\phi + (2kl + mg)l\,sin\phi = -(c/2)\dot{\phi} \quad (1)$$

where l, m are the length of the rod and the mass of flyball, respectively. r and ϕ represent the distance between the rotational axis and the suspension joint, and the angle between the rotational axis and the rod. The net torque for the rotational machine is the difference between the torque Q produced by the engine and the load torque Q_L, directs the second differential equation of motion for the system and described in (2) as:

$$J\dot{\omega} = Q - Q_L = \gamma\,cos\phi - \beta \quad (2)$$

where γ is a proportionally constant and β is an equivalent torque of the load. The governor is geared directly to the output shaft such that its speed of rotation is proportional to the engine speed, i.e. $\eta = n\omega$. Assuming change in time scale $\tau = \Omega_n t$, the equations (1) and (2) of the rotational machine with flyball governor system can be described in nonautonomous form as follows.

$$\begin{cases} \dot{\phi} = \varphi \\ \dot{\varphi} = d\omega^2\,cos\phi + (e + p\omega^2)\,sin\phi\,cos\phi - sin\phi - b\varphi \\ \dot{\omega} = q cos\phi - F \end{cases} \quad (3)$$

where the dot (.) presents the derivative with respect to τ. Parameters are $d = \frac{n^2 mr}{2kl+mg}$, $e = \frac{2kl}{2kl+mg}$, $p = \frac{n^2 ml}{2kl+mg}$, $b = \frac{c}{2ml^2\Omega_n}$, $q = \frac{\gamma}{J\Omega_n}$, $F = \frac{\beta}{J\Omega_n}$ and $\Omega_n = \sqrt{\frac{2kl+mg}{ml}}$. When the load torque is varying, i.e. the load torque is not constant but is represented by a constant term and a harmonic term as $(F + a\,cos\,\bar{\omega}t)$, where F, a, $\bar{\omega}$ are constants. Revising the variables as $\phi = x$, $\dot{\phi} = y$ and $\omega = z$ results the dynamics of nonautonomous system in (4).

$$\begin{cases} \dot{x} = y \\ \dot{y} = dz^2\,cos x + \frac{1}{2}(e + pz^2)sin2x - sinx - by \\ \dot{z} = q\,cos x - F - a\,sin\bar{\omega}t \end{cases} \quad (4)$$

The system (4) shows chaotic behaviour for the parameters values $d = 0.08$, $e = 0.8$, $p = 0.04$, $F = 1.942$, $a = 0.6$, $b = 0.4$, $\bar{\omega} = 1$. The phase plane behaviour is shown in the Fig. 2 and the corresponding time series behaviours are shown in the Fig. 3. The initial condition is chosen as $[0.25;\ 0;\ 0.25]$. Further, the bifurcation is analysed for the parameters q and $\bar{\omega}$. The bifurcation diagrams are shown in the Figs. 4 and 5. Figure 4 shows the bifurcation plot for the parameter q ranges between 0 to 12. It is apparent from the Fig. 4 that the chaotic behaviour exists when the parameter ranges between $2.5 < q < 11.5$ except there exist a periodic window for parameter $4.4 \leq q \leq 4.8$, period doubling and period-3 bifurcations lie for the parameter $6 \leq q \leq 6.2$ approximately. Figure 5 shows the bifurcation plot for the parameter $\bar{\omega}$ ranges between 0 to 2 and reflects the chaotic nature when the parameter ranges between $0.49 < \bar{\omega} \leq 1.24$ and period-1 window ranges for the parameter values $0 \leq \bar{\omega} \leq 0.49$.

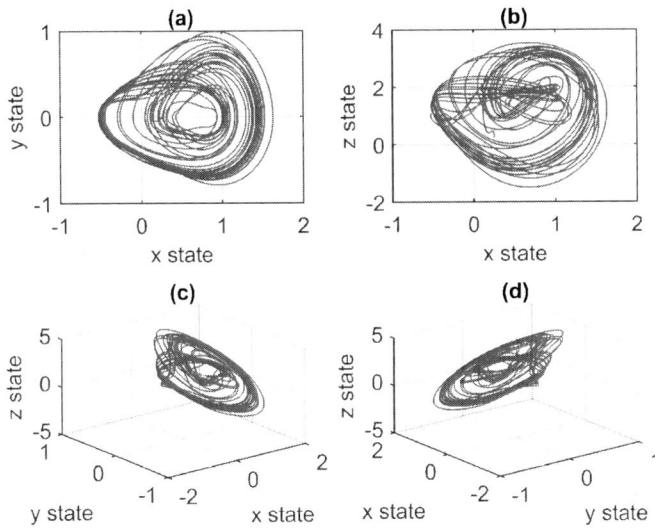

Figure 2. Phase plane behaviour of the centrifugal flyball governor system.

The control aspect of such type of chaos behaviour in the rotational machine such as centrifugal flyball governor system is required because the proper chaos control may bring the new diagnostic tool to prevent something bad or undesirable situation under the operational mode of the rotational machine.

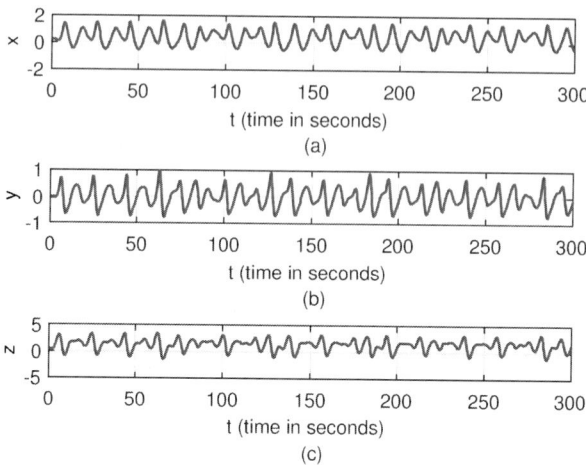

Figure 3. Time series plot shows the chaotic behaviour of the centrifugal flyball governor system.

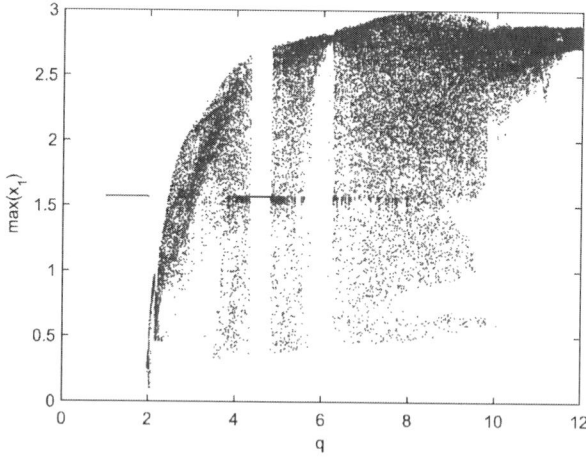

Figure 4. Shows the bifurcation diagram of the centrifugal flyball governor system for parameter q.

Therefore, motivated with the above severe issues, this paper aims at chaos control of the centrifugal flyball governor system. A recursive back-

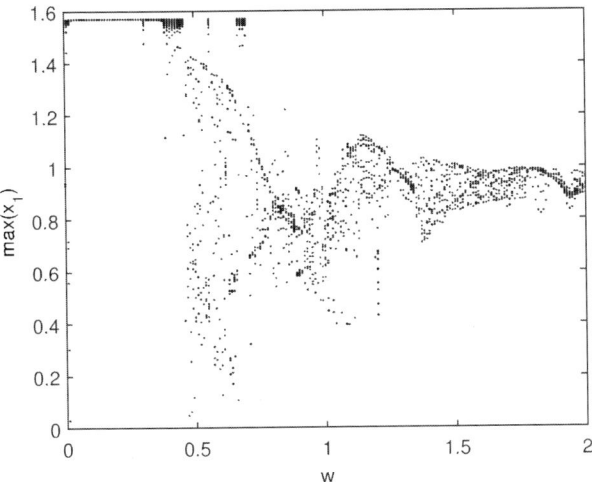

Figure 5. Bifurcation diagram of the centrifugal flyball governor system for parameter $\bar{\omega}$.

stepping sliding mode control (RBSMC) technique is proposed to control the chaotic behaviour of the centrifugal flyball governor system. The general description about the proposed control design is discussed in the next section.

3. GENERAL DESCRIPTION ON PROPOSED RBSMC TECHNIQUE

Consider the general nonlinear system as

$$\begin{cases} \dot{x}_i = f_i(x_i) + b_i(x_i)x_{i+1} \\ \dot{x}_n = f_n(x_n) + b_n(x_n)u \\ y = x_1 \end{cases} \quad (5)$$

where $x_i = [x_1, x_2, ..., x_n]^T$, $f_i(x_i) \to 0$ when $x_i \to 0$ ($x_i \in \Re$), $b_i(x_i) \neq 0$ $i \in 1, 2, ..., n$. u is the required control input to be designed such that the states x_i can track the desired states $x_{id} = f(t)$.

The recursive backstepping sliding mode control (RBSMC) design process is as follows.

Step 1: Consider the control goal $y = x_1 \rightarrow x_{id}$, take x_2 as the first virtual control and pretend that x_{2d} satisfies the following equation.

$$\dot{x}_1 = f_1(x_1) + b_1(x_1)x_{2d} \qquad (6)$$

Define the first tracking error as

$$e_1 = x_1 - x_{1d} = x_1 - y_t \qquad (7)$$

Let the error e_1 be converged to zero exponentially, therefore select the desired behaviour to be $\dot{e}_1 = -\lambda_2 e_1$ and x_{2d} is chosen to satisfy the required dynamic characteristic.

$$\dot{e}_1 = \dot{x}_1 - \dot{x}_{1d} = f_1(x_1) + b_1(x_1)x_{2d} - \dot{x}_{1d} = -\lambda_1 e_1 \qquad (8)$$

Then the following equality corresponds to the desired behaviour as

$$x_{2d} = \frac{1}{b_1(x_1)}(-\lambda_1 e_1 - f_1(x_1) + \dot{x}_{1d}) \qquad (9)$$

Step 2: but we cannot just choose x_2 to be x_{2d}, so we "step back" one integrator to the $\dot{x}_2$ equation. Choose x_3 as the second virtual control to solve such that $x_2 \rightarrow x_{2d}$ as tracking problem. Introduce

$$\dot{x}_2 = f_2(x_2) + b_2(x_2)x_{3d} \qquad (10)$$

Define the first tracking error as

$$e_2 = x_2 - x_{2d} \qquad (11)$$

Let the error e_2 be converged to zero exponentially. Then select the desired behaviour to be $\dot{e}_2 = -\lambda_2 e_2$. x_{3d} is chosen to satisfy the required dynamic characteristic.

$$\dot{e}_2 = \dot{x}_2 - \dot{x}_{2d} = f_2(x_2) + b_2(x_2)x_{3d} - \dot{x}_{2d} = -\lambda_2 e_2 \qquad (12)$$

Then the following equality corresponds to desired behaviour as

$$x_{3d} = \frac{1}{b_2(x_2)}(-\lambda_2 e_2 - f_2(x_2) + \dot{x}_{2d}) \qquad (13)$$

Step i: Choose x_{i+1} as the ith virtual control to solve the $x_i \rightarrow x_{id}$ tracking problem. Define the ith tracking error as

$$e_i = x_i - x_{id} \tag{14}$$

Select the desired behaviour to be $\dot{e}_i = -\lambda_i e_i$. $x_{(i+1)d}$ is chosen to satisfy the required dynamic characteristic.

$$\dot{e}_i = \dot{x}_i - \dot{x}_{id} = f_i(x_i) + b_i(x_i)x_{(i+1)d} - \dot{x}_{id} = -\lambda_i e_i \tag{15}$$

Then the following equality corresponds to desired behaviour as

$$x_{(i+1)d} = \frac{1}{b_i(x_i)}(-\lambda_i e_i - f_i(x_i) + \dot{x}_{id}) \tag{16}$$

Step n: Choose u to solve the $x_n \rightarrow x_{nd}$ tracking problem. Define the nth tracking error as

$$e_i = x_i - x_{id} \tag{17}$$

The error dynamics is written as:

$$\dot{e}_n = \dot{x}_n - \dot{x}_{nd} = f_n(x_n) + b_n(x_n)u - \dot{x}_{nd} \tag{18}$$

Now, we need to design a SMC law $u(t)$ to achieve the asymptotic stability of the error dynamics. The SMC technique involves two steps. First step is to select appropriate sliding surfaces which guarantee the stability of the equivalent error dynamics in the sliding mode such that error (18) converges to zero. Let the proportional integral sliding surface $s(t)$ be defined as:

$$s_n = e_n + \int_0^t \lambda_n e_n \, d\tau \tag{19}$$

The system operates in sliding mode when it satisfies $\dot{s}_n = 0$, i.e.

$$\dot{s}_n = \dot{e}_n - \lambda_n e_n = 0 \tag{20}$$

Error e_n converges to zero exponentially, when it satisfies (20). To establish appropriate sliding surface (19), the second step is to design an SMC to guarantee that the error trajectories reach to the sliding surface $s_n = 0$. Therefore, to ensure the existence of the sliding motion, the recursive backstepping sliding mode control law is proposed as follows:

$$u = \frac{1}{b_n(x_n)}[-f_n(x_n) + \dot{x}_{nd} - \lambda_n e_n - \psi \, sgn(s_n)] \tag{21}$$

Then, the error dynamics is deduced as

$$\begin{cases} \dot{e}_1 = -\lambda_1 e_1 + b_1(x_1)e_2 \\ \dot{e}_2 = -\lambda_2 e_2 + b_2(x_2)e_3 \\ \vdots \\ \dot{e}_{n-1} = -\lambda_{n-1} e_{n-1} + b_{n-1}(x_{n-1})e_n \\ \dot{e}_n = -\lambda_n e_n - \psi\, sgn(s_n) \end{cases} \quad (22)$$

Proposition 1: When the parameters $\lambda_i > 0$ and $b_i(x_i) = c_i$ for $i = 1, 2, ..., n$, where c_i is positive constants. It is obvious that the error e_i converge to origin exponentially, i.e. $x_i \to x_{id}$.

Proposition 2: When $b_i(x_i) = c_i$ is the function of x_i and we select the parameters $\lambda_i > 0$ for $i = 1, 2, ..., n$, it can make error e_i globally stable at origin, i.e. $x_i \to x_{id}$.

Proof: Let a Lyapunov function candidate be defined as:

$$V(s_n) = \frac{1}{2} s_n^2 \quad (23)$$

Performing the first order partial derivative of Lyapunov function candidate (23) and using (18),

$$\dot{V}(s_n) = s_n \dot{s}_n = s_n [f_n(x_n) + b_n(x_n)u - \dot{x}_{nd}] \quad (24)$$

By using the error dynamics (22) and the control law (21), we get

$$\dot{V}(s_n) \leq -\psi |s_n| \quad (25)$$

One can show that $\dot{V}(s_n) < 0$ when $s_n(t) \neq 0$ by suitable choice of $\psi > 0$. Thus, according to Lyapunov stability theory, $s_n(t)$ always converges to sliding surface $s_n(t) = 0$. Therefore, the error dynamics in sliding manifold is asymptotically stable, i.e., the error dynamics converges to zero. ,

4. RECURSIVE BACKSTEPPING SLIDING MODE CONTROL OF CHAOTIC CENTRIFUGAL FLYBALL GOVERNOR SYSTEM

The centrifugal flyball governor system (4) with added control input $u(t)$ is rewritten as:

$$\begin{cases} \dot{x}_1 = x_2 \\ \dot{x}_2 = dx_3^2 \, cosx_1 + \frac{1}{2}(e + px_3^2) \, sin2x_1 - sinx_1 - bx_2 \\ \dot{x}_3 = q \, cosx_1 - F - a \, sin\bar{\omega}t + u(t) \end{cases} \quad (26)$$

Our objective is to design the control $u(t)$ such that the states x_1, x_2 and x_3 can track the desired states $x_{1d} = f(t)$. Let the error be defined as:

$$\begin{cases} e_1 = x_1 - x_{1d} \\ e_2 = x_2 - \lambda_1 e_1 \\ e_3 = x_3 - \lambda_2 e_2 \end{cases} \quad (27)$$

x_{2d} and x_{3d} are defined as $\lambda_1 e_1$ and $\lambda_2 e_2$, respectively, where λ_1, λ_2 are control parameters tuned by the designer. Then the error dynamics is written as:

$$\begin{cases} \dot{e}_1 = e_2 + \lambda_1 e_1 - \dot{f}(t) \\ \dot{e}_2 = zx_3^2 cos(e_1 + f(t)) + \frac{1}{2}(e + px_3^2) \, sin2(e_1 + f(t)) - sin(e_1 + f(t)) \\ \quad -b(e_2 + \lambda_1 e_1) \\ \dot{e}_3 = q \, cos(e_1 + f(t)) - F - a \, sin\bar{\omega}t + u(t) \end{cases} \quad (28)$$

Now, we need to design a SMC law $u(t)$ to achieve the asymptotic stability of the error dynamics (28). Let the proportional integral sliding surface $s(t)$ be defined as:

$$s(t) = e_3 + \int_0^t (\lambda_2 e_2 + \lambda_1 e_1) \, d\tau \quad (29)$$

The control law $u(t)$ is designed in (30) to achieve the occurrence of the sliding motion.

$$u(t) = (F + a \, sin\bar{\omega}t) - q \, cos(e_1 + f(t)) - (\lambda_2 e_2 + \lambda_1 e_1) - \psi sgn(s) \quad (30)$$

Control law $u(t)$ in (30) ensures the trajectories of the biological system (3) converge to the sliding surface $s(t) = 0$. As a proof, let a Lyapunov function candidate be defined as:

$$V(s) = \frac{1}{2}s^2 \tag{31}$$

Performing the first order partial derivative of Lyapunov function candidate (31) and using (28),

$$\dot{V}(s) = s\dot{s}$$
$$= s[\dot{e}_3 + \lambda_2 e_2 + \lambda_1 e_1]$$
$$= s[q\cos(e_1 + f(t)) - F - a\sin\bar{\omega}t + \lambda_2 e_2 + \lambda_1 e_1 + u(t)]$$

By using the error dynamics (28) and the control law (30), we get

$$\dot{V}(s) \leq -\psi|s| \tag{32}$$

One can show that $\dot{V}(s) < 0$ when $s(t) \neq 0$ by suitable choice of $\psi > 0$. Thus, according to Lyapunov stability theory, $s(t)$ always converges to sliding surface $s(t) = 0$. Therefore, the error dynamics in sliding manifold is asymptotically stable, i.e. the error dynamics converges to zero.

The designed control law (30) is summarized in (33) when the desired state is $f(t) = 0$:

$$u(t) = (F + a\sin\bar{\omega}t) - q\cos(e_1) - (\lambda_2 e_2 + \lambda_1 e_1) - \psi sgn(s) \tag{33}$$

and when the desired state $f(t)$ is chaotic, the control law is written in (30). In the next section, results are presented to validate the proposed approach and discussion is summarized.

5. Results and Discussion

MATLAB ode45 is used for the simulation. Simulation runs for 100 seconds with step time 0.001. The initial condition of the system for the dynamic desired behavior is considered as $[x_{1d}(0) \ x_{2d}(0) \ x_{3d}(0)] = [1 \ 1 \ 1]$. The parameters values of the system (4) is considered as $d = 0.08, e = 0.8, p = 0.04, F = 1.942, a = 0.6, b = 0.4, \bar{\omega} = 1$ and $\lambda_1 = \lambda_2 = 2$. The controlled response corresponding to the desired dynamic behaviours are

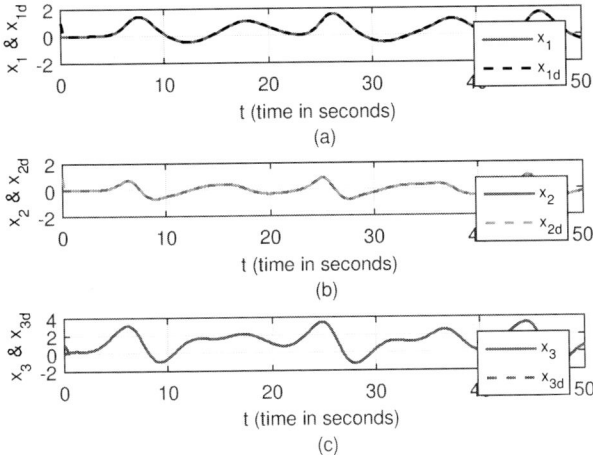

Figure 6. Controlled and desired states of the flyball governor system.

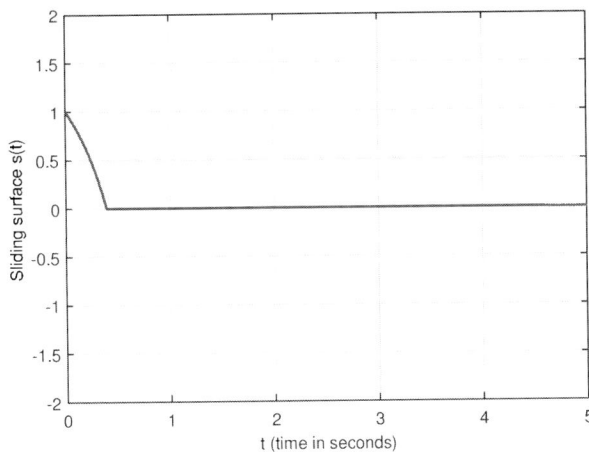

Figure 7. Shows the designed sliding surface.

shown in the Fig. 6. Some of the figures are restricted to 5 second for better and clear visualization of the depicted behaviours. The designed sliding surface is depicted in the Fig. 7 which reflects the switching nature of the surface. The required control input is shown in the Fig. 8. The peak to peak

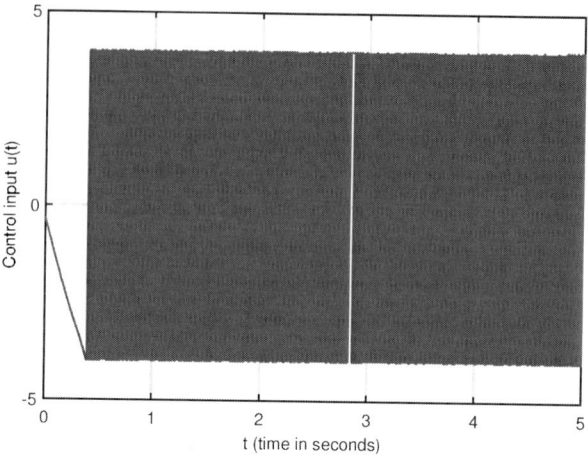

Figure 8. Designed control input to achieve the tracking of the desired behaviour.

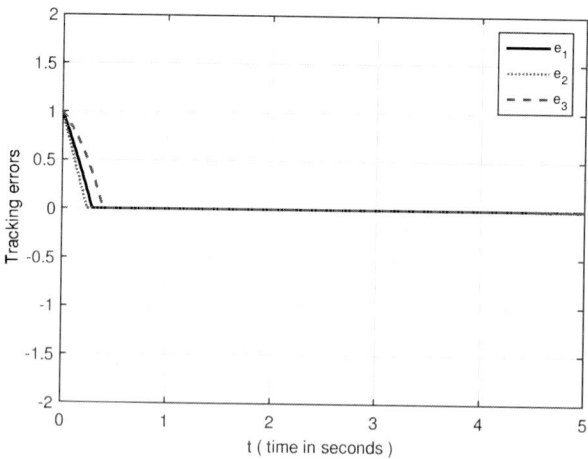

Figure 9. Shows the tracking errors between the original and the desired states.

amplitude of the designed control input is acceptable and implementable in the real time scenario. The tracking errors are shown in the Fig. 9. The Figs. 6 and 9 reveal that the transient time is very less the steady state is achieved in quick time, i.e. approximately less than 0.45 second.

Similarly, the proposed controller can stabilize the chaotic dynamics of the rotational machine with centrifugal flyball governor system to the desired set point $f(t) = 0$ and corresponding result is shown in the Fig. 10.

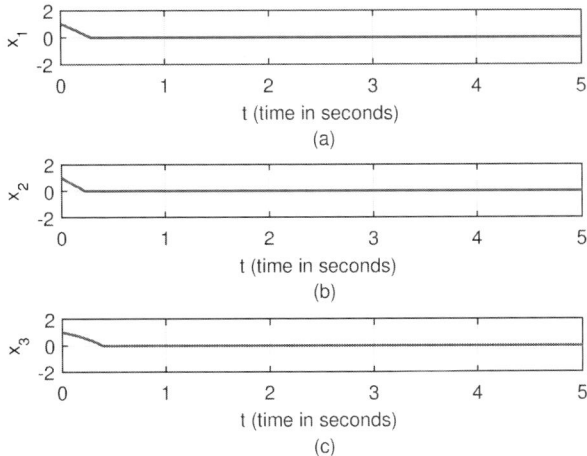

Figure 10. Controlled states of the flyball governor system when the desired set point is $f(t) = 0$.

CONCLUSION AND FUTURE SCOPE

In this paper, chaos control of a rotational machine with centrifugal flyball governor system is achieved. The control of chaos is achieved by designing a new controller. A recursive backstepping sliding mode control (RBSMC) is designed such that it can control the chaotic dynamics in the rotational machine with centrifugal governor system to stabilize at any position or to track any desired trajectory that is a smooth function of time. Here, the backstepping technique based on sliding mode error is designed to adopts backstep design process. A proportional integral switching surface based on errors is designed to the backstepping control design as well as to achieve the stability condition of the error dynamics. The design of controller depends on the corresponding proportional integral sliding mode errors. The proposed controller does not require the parameter perturbation. Lyapunov stability theory is used to derive the required stability condition. MATLAB environment is used for numerical simulation. Simulation results reveal the successful achievement of the objective, i.e. chaos control of a rotational machine with centrifugal flyball governor system using recursive backstepping sliding mode control technique is achieved.

Finally, in order to have more access to the practical application, analysis and control of the time-delayed and fractional-order centrifugal flyball governor systems will be quite challenging aspects for the researchers in the future.

REFERENCES

[1] Slotine J. E., W. Li, *Applied Nonlinear Control*, Prentice Hall, New Jersey, 1991.

[2] Wiggins S., *Introduction to Applied Nonlinear Dynamical Systems and Chaos*, Springer, New York, 1990.

[3] Wiggins S., *Global Bifurcations and Chaos*, Springer, New York, 1988.

[4] Zhang H., Chen D., Xu B., Patelli E., Tolo S., Dynamic analysis of a pumped-storage hydropower plant with random power load, *Mechanical Systems and Signal Processing* 100 (2018) 524–533. doi:10.1016/j.ymssp.2017.07.052.

[5] Zhang H., Chen D., Wu C., Wang X., Dynamics analysis of the fast-slow hydroturbine governing system with different time-scale coupling, *Commun. Nonlinear Sci. Numer. Simulat.* 54 (2018) 136–147. doi:10.1016/j.cnsns.2017.05.020.

[6] Xu B., Chen D., Zhang H., Wang F., Zhang X., Wu Y., Hamiltonian model and dynamic analyses for a hydro-turbine governing system with fractional item and time-lag, *Commun Nonlinear Sci. Numer. Simulat.* 47 (2017) 35–47. doi:10.1016/j.cnsns.2016.11.006.

[7] Ge Z. M., Lee C., Non-linear dynamics and control of chaos for a rotational machine with a hexagonal centrifugal governor with a spring, *Journal of Sound and Vibration* 262 (2003) 845–864. doi:10.1016/S0022-460X(03)00092-0.

[8] Ge Z. M., Lee C., Anticontrol and synchronization of chaos for an autonomous rotational machine system with a hexagonal centrifugal governor, *Journal of Sound and Vibration* 282 (2005) 635–648. doi:10.1016/j.jsv.2004.03.034.

[9] Ge Z. M., Lee C., Control, anticontrol and synchronization of chaos for an autonomous rotational machine system with time-delay, *Chaos, Solitons and Fractals* 23 (2005) 1855–1864. doi:10.1016/j.chaos.2004.07.023.

[10] Zhang J. G., Yu J. N., Chu Y. D., Li X. F., Chang Y. X., Bifurcation and chaos of a non-autonomous rotational machine systems, *Simulation Modelling Practice and Theory* 16 (2008) 1588–1605. doi:10.1016/j.simpat.2007.09.009.

[11] Lin Q., Wu X., Gui Z., A global synchronization criterion for the chaotic rotational machine systems under linear state error feedback control, in: Second International Conference on Intelligent Computation Technology and Automation, 2009. doi:10.1109/ICICTA.2009.181.

[12] Zhang J. G., Li X. F., Chu Y. D., Yu J. N., Chang Y. X., Hopf bifurcations, lyapunov exponents and control of chaos for a class of centrifugal flywheel governor system, *Chaos, Solitons and Fractals* 39 (2009) 2150–2168. doi:10.1016/j.chaos.2007.06.131.

[13] Zhang J. G., Mello L. F., Chu Y. D., Li X. F., An X. L., Hopf bifurcation in an hexagonal governor system with a spring, *Commun Nonlinear Sci Numer Simulat* 15 (2010) 778–786. doi:10.1016/j.cnsns.2009.04.035.

[14] Wang J., Wang H., Guo L., Anti-control of chaos in mechanical centrifugal governor system, *Journal of Information and Computational Science* 9 (2012) 5127–5133. doi:http://www.joics.com.

[15] Aghababa M. P., Design of an adaptive finite-time controller for synchronization of two identical/different non-autonomous chaotic flywheel governor systems, *Chin. Phys. B* 21 (2012) 030502. doi:10.1088/1674-1056/21/3/030502.

[16] Aghababa M. P., Aghababa H. P., Finite-time stabilization of non-autonomous uncertain chaotic centrifugal flywheel governor systems with input nonlinearities, *Journal of Vibration and Control* 19 (2012) 1–10. doi:10.1177/1077546312463715.

[17] Peng J. K., Yu J. N., Zhang L., Hu P., Study on synchronization of the centrifugal flywheel governor system, *Applied Mechanics and Materials: Advances in Mechatronics and Control Engineering II* 433–435 (2013) 21–29. doi:10.4028/www.scientific.net/AMM.433-435.21.

[18] Raghavendra G. H., Kumarappa S., Retrofication of mechanical speed governor with electronic speed governor for heavy duty diesel engines, *International Journal for Research in Applied Science and Engineering Technology* 3 (2015) 383–388. doi:www.ijraset.com.

[19] Luo S., Hou Z., Zhang T., Performance enhanced design of chaos controller for the mechanical centrifugal flywheel governor system via adaptive dynamic surface control, *AIP Advances* 6 (2016) 095217. doi:10.1063/1.4963645.

[20] Song J., Sun K., Ling S., Stabilization and synchronization for a mechanical system via adaptive sliding mode control, *ISA Transactions* 68 (2017) 353–366. doi:10.1016/j.isatra.2017.02.013.

[21] Rao X. B., Chu Y. D., Chang Y. X., Zhang J. G., Broken farey tree and fractal in a hexagonal centrifugal governor with a spring, *Chaos, Solitons and Fractals* 107 (2018) 251–255. doi:10.1016/j.chaos.2018.01.015.

[22] Singh P. P., Singh J. P., Roy B. K., Synchronization and anti-synchronization of Lu and Bhalekar-Gejji chaotic systems using nonlinear active control, *Chaos, Solitons and Fractals* 69 (2014) 31 – 39. doi:10.1016/j.chaos.2014.09.005.

[23] Singh P. P., Singh J. P., Roy B. K., Synchronization of chaotic systems using nac and its application to secure communication, *Int. J. Control Theory and Application* 8 (2015) 994 – 1003.

[24] Borah M., Singh P. P., Roy B. K., Improved chaotic dynamics of a fractional-order system, its chaos-suppressed synchronisation and circuit implementation, *Circuits, Systems, and Signal Processing* 35 (6) (2016) 1871–1907. doi:10.1007/s00034-016-0276-9.

[25] Singh P. P., Singh J. P., Roy B. K., Implementation for synchronization of Chen and Liu-Yang chaotic systems using SMC and active control schemes, *Int. J. Control Theory and Application* 9 (2016) 159 – 169.

[26] Singh P. P., Singh J. P., Roy B. K., SMC based synchronization and anti- synchronization of chaotic systems for secure communication and analog circuit realization, *Int. J. Control Theory and Application* 9 (2016) 171 – 183.

[27] Singh P. P., Singh J. P., Roy B. K., NAC-based synchronisation and anti-synchronisation between hyperchaotic and chaotic systems, its analogue circuit design and application, *IETE Journal of Research* 63 (2017) 1 – 17. doi:10.1080/03772063.2017.1331758.

[28] Singh P. P., Roy B. K., Comparative performances of synchronisation between different classes of chaotic systems using three control techniques, *Annual Reviews in Control* 45 (2018) 152 – 165. doi:10.1016/j.arcontrol.2018.03.003.

ABOUT THE EDITOR

Dr. Piyush Pratap Singh earned his Bachelor of Technology degree in Electronics and Communication Engineering from Institute of Engineering and Technology, Dr. Ram Manohar Lohia Avadh University Faizabad, Uttar Pradesh, India in the year 2008. He joined as a lecturer in the Department of Electronics and Communication Engineering at LIET Alwar, Rajasthan, India. In the year 2012, he earned his Master of Technology degree with the specialization Signal Processing and Control in the department of Electrical Engineering at National Institute of Technology Hamirpur, Himachal Pradesh, India. In the year 2016, he received his Doctor of Philosophy degree in the specialization Control and Industrial Automation from the Department of Electrical Engineering, National Institute of Technology Silchar, Cachar, Assam, India. In 2016, he was appointed as an Assistant Professor in the Department of Electrical Engineering at National Institute of Technology Meghalaya, Shillong, Meghalaya, India and since then continued his professional carrier. He has approximately six years of teaching and research experience wherein he taught several courses at the undergraduate level, namely Basic Electrical, Network Theory, Signals & Systems, Analog Electronics, Control Systems, and Advanced Control Systems. He also taught some courses at the postgraduate level named Instrumentation and Control in Energy Systems and Advanced Control Systems. He has been engaged in administrative activities and is a member and convener of different standing committees and Professor in Charge of the Institute Security Committee. His research interest includes control systems, advanced control, chaos synchronization, chaos control, identification of chaos in different field of science and electrical engineering, information technology, communication engineering and mechanical engineering, synchronization in complex networks etc. He has published more than fifty research papers in reputed international and national journals and conferences including book chapters and books. Also, he is actively engaged in various professional and outreach activities such as being a senior member of IEEE society, life member of IE society and member of ACDOS society.

Email: piyushpratapsingh@gmail.com

INDEX

A

adaptive, 9, 21, 26, 50, 81, 84, 85, 86, 109, 110, 114, 115, 116, 131
  control, vii, ix, 2, 7, 19, 20, 21, 22, 23, 25, 26, 27, 29, 30, 32, 35, 36, 38, 39, 40, 45, 46, 47, 48, 49, 50, 51, 54, 55, 56, 57, 59, 60, 63, 64, 65, 67, 68, 69, 70, 72, 74, 75, 77, 80, 81, 83, 84, 85, 86, 87, 88, 89, 91, 92, 93, 94, 95, 96, 101, 104, 106, 107, 108, 109, 110, 111, 113, 114, 115, 116, 117, 119, 120, 121, 122, 123, 124, 125, 126, 127, 128, 129, 130, 131, 132, 133
applications, ix, 19, 21, 28, 29, 46, 47, 109, 110, 111, 116

B

backstepping, vii, ix, 50, 84, 85, 110, 113, 114, 116, 117, 122, 129, 130
backstepping technique, 116, 129
bifurcation diagram, 3, 6, 7, 119, 120
bifurcation(s), 3, 6, 7, 19, 20, 46, 84, 85, 89, 90, 91, 109, 115, 119, 120, 121, 130, 131

C

chaos, vii, ix, x, 2, 3, 9, 19, 20, 21, 22, 24, 25, 26, 27, 28, 29, 33, 37, 46, 47, 48, 49, 51, 80, 81, 83, 84, 85, 86, 87, 89, 108, 109, 110, 111, 113, 114, 115, 116, 117, 119, 120, 121, 123, 125, 127, 129, 130, 131, 132, 133
chaos synchronization, 114, 115, 116, 130, 131, 132, 133
chaotic, v, vii, ix, 1, 2, 3, 4, 5, 6, 7, 8, 9, 10, 11, 12, 13, 14, 17, 18, 19, 20, 21, 22, 24, 25, 26, 27, 28, 29, 30, 33, 34, 41, 46, 47, 48, 53, 57, 65, 67, 84, 85, 86, 87, 89, 90, 91, 100, 101, 102, 103, 104, 109, 110, 111, 113, 114, 115, 116, 117, 118, 119, 120, 121, 126, 129, 131, 132
  circuits, 20
  oscillations, 84, 85, 109
chaotic behavior, 20, 21, 47
chaotic masking, 9, 25, 26, 28, 48
chattering, 49, 51, 58, 80, 86, 108, 110
coupling, 9, 26, 130

D

data communication, 21, 24, 47
decoding, 2, 24

Index

dynamics, ix, 1, 3, 8, 9, 12, 13, 16, 19, 20, 25, 29, 31, 33, 36, 38, 40, 46, 47, 49, 50, 51, 52, 53, 54, 55, 56, 58, 59, 62, 67, 68, 72, 81, 83, 85, 88, 89, 91, 92, 93, 95, 96, 109, 110, 113, 114, 118, 123, 124, 125, 126, 129, 130, 132

E

eigenvalues, 11, 16
encoding, 2, 24
encryption, ix, 2, 8, 19, 23, 24, 27, 28, 29, 30, 33, 40, 42, 45

F

fractional-order, 84, 85, 110
fuzzy, 50, 81, 85
fuzzy logic, 85
fuzzy logic control, 85

H

hydraulic, 110
hyperchaotic systems, vii, 7, 8, 23, 25, 30, 32, 33, 34, 36, 40, 45

I

integral, 50, 53, 54, 59, 62, 67, 72, 84, 86, 91, 93, 95, 107, 108, 110, 111, 113, 123, 125, 130

J

Jacobian matrix, 10, 16

L

limit cycle, 20
Liu, 20, 21, 46, 47, 81, 109, 110, 132
Lorenz system, 19, 20, 46, 47
Lu system, 19, 46
Lyapunov, 1, 3, 4, 5, 6, 7, 11, 12, 17, 18, 19, 23, 32, 33, 36, 38, 40, 45, 54, 55, 60, 63, 70, 72, 83, 91, 97, 109, 114, 115, 116, 124, 126, 130
  exponents, 1, 4, 7, 11, 12, 17, 91, 115
  function, 32, 36, 40, 55, 60, 63, 70, 72, 97, 115, 116, 124, 126
  spectrum, 5, 6, 11, 12, 17, 18
Lyapunov function, 32, 36, 40, 55, 60, 63, 70, 72, 97, 115, 116, 124, 126

M

Memristor, 22, 48

N

neural network, 81, 115
nonlinear dynamic systems, 24, 84
nonlinear dynamics, ix, 109
nonlinear systems, 84

O

oscillation, 84, 85, 86, 87, 103, 115
oscillators, 8, 29, 110
outreach, 134

P

proportional, 50, 53, 54, 57, 67, 84, 86, 91, 93, 107, 108, 110, 111, 113, 118, 123, 125, 130

Q

quality improvement, 111

R

response, 25, 26, 27, 30, 31, 32, 33, 34, 35, 36, 38, 39, 40, 41, 42, 43, 44, 57, 58, 65, 66, 74, 77, 78, 86, 94, 100, 101, 102, 103, 104, 105, 126
rotational inertia, 88

Index

S

secure communication, ix, 8, 20, 21, 23, 24, 25, 29, 33, 35, 37, 45, 47, 48, 132
sliding mode, vii, ix, 49, 50, 51, 53, 54, 58, 59, 62, 65, 66, 68, 71, 72, 74, 76, 77, 79, 80, 81, 83, 85, 86, 87, 92, 93, 95, 96, 102, 107, 108, 110, 111, 113, 116, 117, 121, 122, 123, 129, 130, 131
sliding mode control, vii, ix, 49, 50, 51, 53, 54, 58, 62, 65, 66, 71, 74, 76, 77, 79, 80, 81, 83, 85, 86, 87, 92, 93, 96, 107, 108, 110, 111, 113, 116, 117, 121, 122, 123, 129, 130, 131
stability analysis, 67, 85, 87
Lyapunov, 23, 60, 83, 124, 126, 130
stabilization, 115, 131
synchronization, 114, 115, 116, 130, 131, 132, 133

T

techniques, ix, 8, 23, 24, 27, 28, 29, 30, 33, 38, 40, 41, 42, 45, 48, 49, 51, 64, 65, 66, 74, 77, 78, 79, 80, 84, 86, 91, 104, 105, 106, 107, 108, 111, 114, 116, 132
theoretical approach, 1, 24, 50, 84
torus, 17, 18, 19

U

unstable, 11, 16, 86

V

variables, ix, 1, 3, 4, 7, 9, 13, 14, 15, 16, 17, 18, 19, 20, 23, 24, 25, 28, 29, 30, 31, 33, 45, 46, 47, 75, 77, 86, 91, 99, 118
variations, 26, 49, 50, 81, 84, 86